Military Aircraft Insignia of the World

Military Aircraft Insignia of the World

John Cochrane and Stuart Elliott

FLIGHT
RECORDER
PUBLICATIONS
A passion for accuracy

First published in Great Britain in 2014 by
Crécy Publishing

ISBN 9780955 426872

Printed in Malta by Melita Press

Crécy Publishing Limited
1a Ringway Trading Estate
Shadowmoss Road
Manchester M22 4LH

www.crecy.co.uk

Once the flying machine was a practical proposition, its military use became obvious, and was, in fact, a continuation of the use that had been made of the balloon. Centuries before the Wright Brothers' success, fictional accounts of war in the air were very popular, some written strictly for adventure, some as dire warnings. The 1899 Hague Convention outlawed all methods of dropping explosives from aerial devices (at that time only balloons, of course), and this was signed by all the major powers. Although balloons had been used to drop explosives, their inability to move freely through the air made them somewhat inappropriate. They did, however, work well as an observation platform, which is how many military minds saw the first aeroplanes. From the very beginning of powered flight, governments and individuals controlled by governments were developing aircraft for this use, and they therefore needed to be marked as government property. Flying a flag was the obvious solution, but this proved to be ineffective and even dangerous. The painting of a representation of a flag was the natural alternative.

The first known use of markings to identify the nationality of aircraft was at the 1910 Bombing Competition in Vienna. Each competing machine carried its national colours as wing-tip stripes. We know that Russia, France, Italy, Romania, Poland and Bohemia took part in this competition, Poland and Bohemia carrying the red and white band of Austria. Prior to this, during the American Civil War, the baskets of the Union balloons were painted with the stars and stripes to distinguish them from the few Confederate balloons.

By the outbreak of the First World War, many countries had already used aircraft for military purposes, usually for reconnaissance but occasionally for dropping bombs. Italy, France, Spain, Bulgaria, Turkey, Serbia, Greece, Romania, Mexico, the USA and even China had made some use of aircraft in military conflicts between 1911 and 1914.

By an order dated 26 July 1912, France became the first country to specify the precise shape, size and colour of military markings for aircraft with a roundel form of the French flag.

This form had been used by some military units as far back as the Napoleonic wars. Romania appeared to be the second country to adopt a roundel format.

The First World War accelerated military aircraft development, and saw the first combat between aircraft. This necessitated a coherent form of national identification, not just for opposing aircraft but also for ground troops, who would wish to know whether overflying machines were friend or foe. By the end of the war the system was well recognised, but there were problems. Many Allied airmen disliked the painting of what looked like a target on the side of their aircraft. All military aviators were uncomfortable with the idea of sitting exactly between two wing markings, again making the man an obvious target. The white parts of insignia negated the principles of camouflage, and at the very end of the war these were eliminated for night use.

There were other uses of roundel-type markings. The stunt pilot B. C. Hucks marked a red roundel on the top of the wings of his aircraft

A very early use of a flag-inspired roundel (but not for military use!) on a Norwegian Bleriot.

to prove to his audience that he had in fact done a 'loop', an adventurous stunt in 1914. The intrepid Norwegian pilot Tryggve Gran made the first crossing of the North Sea between Scotland and Norway in July 1914; to prove that he had made the trip he painted a roundel form of the Norwegian flag on the wings of his Bleriot.

During the Second World War white areas were again reduced or eliminated, and in the Pacific area red parts of insignia were painted out owing to possible confusion with Japanese markings. Neutral countries, which felt the need to underline their position, used much bolder versions of their insignia.

With the dramatic increase in the speed of aircraft since the 1970s, national markings have become almost totally irrelevant. In combat situations many nations now use toned-down black or grey versions of their national insignia. Although these modifications were first seen in the 1920s, the modern version was to confuse advanced missiles, which targeted bright areas on aircraft, as well as being less visible on current radar equipment. In some guerrilla wars in southern Africa in the 1970s and 1980s national markings were dispensed with altogether.

So we have come almost full circle again as in the slower-paced Third World national markings have a very important part to play, if only because possession of an air force is a required status symbol in many smaller countries. Many of these countries have used aircraft provided by, maintained by, and flown by mercenaries from one of the major powers. For political reasons the national marking of the country involved has been used rather than the country of ownership of the aircraft or the nationality of the pilot.

Dates stated herein may be approximate as operational use often lags far behind official instruction.

ACKNOWLEDGEMENTS

A book such as this could not be done without the large number of people from around the world who have willingly provided photos and information and especially encouragement, many being representatives of air forces, manufacturers and museums. We are also grateful for the unfailing enthusiasm of Jim Sanders of the Small Air Forces Clearing House, the Air Combat Information Group, and the many branches worldwide of the International Plastic Modellers Society. We would wish to acknowledge especially Greg Kozak, Kevin Curtis, Roberto Gentili, Tom Cooper and Yefim Gordon, among many others too numerous to mention, who have deluged us with information, photographs and encouragement, and of course the team at Crécy Publishing. We thank you all.

ABKHAZIA

This breakaway region of Georgia declared its independence on 23 July 1992. Its few aircraft carried the national flag. The country's flag features a hand (an ancient symbol of the Abkhazian people since the 13th century) and seven stripes and stars (the number seven is part of their tradition). Since 2008 a roundel version of the national flag has been used. South Ossetia is another separatist part of Georgia but does not seem to have any aircraft.

An Abkhazian Air Force Mil-24 in 2010. via Greg Kozak

1924-29

1937-67

1967-79

AFGHANISTAN

The first military aircraft arrived in Afghanistan from Russia in 1921, but the air arm was not established until 22 August 1924. Its few aircraft were destroyed in the revolution of 1929 and a new force was not reformed until 1937.

During the early period the aircraft, mostly of Russian origin, bore the Muslim-based arms of Afghanistan in black on white, and the legend 'Allah u Akbar' ('God is great') below the wings together with the Afghan flag and coat of arms. Mohamed Nadir Shah, the victor in the 1929 insurrection, changed Afghanistan's colours from largely red to red, green and black, symbolising bloodshed for

A Royal Afghan Air Force Hawker Hind, ex-RAF India 1938, now at RAF Cosford. Author

independence, hope for the future, and the country's dark past. These colours were carried as rudder stripes. There is some evidence of a four-colour roundel, black, red and green with a white centre, the central spot being inscribed with the phrase 'Allah u Akbar'. The use of this marking has no photographic confirmation and its use is very unlikely.

On the reorganisation of the air arm in 1937, aircraft were released from British stocks in India, and the RAF roundels were over-painted in Afghan colours. Additionally some aircraft may have sported red, green and black stripes across the wings.

The Royal Afghan Air Force was formed in 1948, and the roundel continued in use until 1967. In the early 1950s the rudder striping gave way to a similarly marked fin flash. In 1967 a new insignia consisting of a three-colour segmented triangle within a white circle was carried on wings, fuselage and fin. The white surround featured the initials of the Afghan armed forces in Arabic script.

A Mil-24, 1982. via Greg Kozak

A Northern Alliance anti-Taliban group MiG 21, 2002. via Greg Kozak

1979-83

A 1988 MiG 21, supplied by the USSR, at Bagram air base. via Greg Kozak

Below: A US-supplied Cessna Grand Caravan, 2010. via Greg Kozak

1983-94

1996

2000

2010

Northern Alliance

The monarchy was overthrown in 1973 and the country fell further within the Soviet sphere of influence. The triangular insignia continued in use until the Russian invasion of 1979, probably with different letters on the roundel. After 1979 a red disc with yellow inscriptions was adopted, and later, in 1983, a red star within a circle of the Afghan national colours.

Following the departure of the Soviet forces in 1989 there was a complete breakdown of the government, and various dissident groups carried their own markings, many of which were discovered during the Allied invasion of 2002. Examples included possible Taliban aircraft, which carried roundels of black, white, green and white, and Hezb-a-Wahdat, using a green disc with a black border with various symbols, or others with similar markings with a central small white spot. The Northern Alliance used the triangle insignia without the red portion.

The three-colour triangle of 1967-79 has been reintroduced with a new inscription, which was changed in 2010 with a change of name to the Afghan National Army Air Force.

1946-49

1958-60

1960-92

1992

ALBANIA

An attempt was made to form an Albanian Air Corps on the country's formation in 1914. This did not come to fruition and, owing to the country's poor economic condition, an air force was not formed until 1947.

Although very much under Soviet influence, the first proposed idea for an aircraft marking was the black two-headed eagle of Albania on a red disc topped with a yellow-outline star; there is, however, no record of its use. The earliest evidence is of a red star on a black disc with a red/black/red fin flash, the black area bearing a red star. The fin marking was first horizontal, later vertical. For a period in the late 1950s a red and black roundel was used, with occasionally a red star on the fin. From about 1960 the usual form of marking was a red and black roundel with a red star on the black inner circle; this was used on wings and fuselage and later on the fin.

In 1993, after the fall of the communist regime, the red star marking was discontinued and a representation of the Albanian national flag was considered. There is no evidence of this being used, and the current marking is a simple red/black/red roundel.

An Albanian Air Force Nanchang CJ6, a Chinese-built version of the Yak-18, 1993. Chris Lofting

ALGERIA

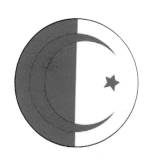

The Armée de l'air Algérienne was formed when the country gained independence in 1962. The markings used were the new Algerian flag in roundel form and the flag as a fin flash. This consisted of two halves, green and white, with a red star and crescent superimposed. For a short period in the early 1960s bars of red, green and white were added on each side of the roundel.

An Algerian Air Force Lockheed C-130H, 2009. Andreas Zeitler

1962-64

ANGOLA

This ex-Portuguese colony gained its independence in 1975 and was immediately plunged into a long and devastating civil war. Military equipment, including modern aircraft, was supplied mostly from the USSR. These aircraft were usually flown by pilots from Cuba and East Germany, but under the Angolan flag.

This flag has always been divided horizontally, red over black with a yellow symbol – the red for liberation, and the black for Africa. The initial marking was simply the rudder striped in red, yellow and black. There is mention of the three colours as roundels and a fin flash comprising the national flag, but there is no confirmation or photographic evidence of these.

From about 1980 the roundel was red outer and black inner with a yellow star on the black, but this was only occasionally used, especially on front-line aircraft.

1975-80

From about 1987 a new roundel was issued in red and black, the colours divided by a wavy line and a yellow star superimposed. Recently the star has been omitted and the roundel given a yellow outline border.

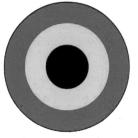

1975

1980

2012

Su-22 carrying the current roundel, without the yellow star.

ARGENTINA

1919 onwards

Navy, 1958-1980

Navy

Argentina's air force was first formed as the Servício Aeronautico Ejercito on 10 August 1912. The white and pale blue colours of the 1816 year of independence were used in roundel form from about 1919. The rudder marking, or fin flash, was a representation of the Argentine flag, which bore on the white section a yellow 'Sun of May', a reference to the sun shining on the morning of the successful liberation from Spain on 25 May 1810.

The navy formed its Comando de Aviacion Naval Argentina on 17 October 1919. This air arm dispensed with the roundel but kept the tail markings. Wings were marked with a large anchor in either black or white, depending on the background colour. On a few aircraft a red anchor across the wings or across the roundel has been noted. The navy and the army kept their separate markings until the formation of the Argentine Air Force on 4 January 1945.

Argentinian Navy Chance-Vought Corsair F4U-5 with special naval insignia, c1957. Santiago Rivas

A Fokker F-28 of the Argentine Air Force. Chris Lofting

In about 1960 a new navy roundel was designed, consisting of a white disc with a thin blue/white/blue border. Within the large white centre was a design comprising a black anchor, a yellow sun and the red 'Cap of Liberty'. This was used frequently, but not always, on the fuselage sides only. Naval aircraft retained the wing anchors and fin flash. Army and air force aircraft continue to use the blue and white roundels, although army helicopters rarely do so, using only the flag.

Some aircraft captured by rebels during the 1955 revolution carried 'V+' and the initials 'MR' crudely painted in red.

Argentina possesses a number of paramilitary organisations that use aircraft. The Gendarmeria Nacional uses normal markings, while the Prefectura Naval uses normal fin flashes and a symbol of two crossed anchors in black or white depending on the background colour.

ARMENIA

Armenia gained independence as the Erivan Republic after the First World War, but its short period of freedom was spent in conflict with Russia, Turkey and its Caucasian neighbours. It is believed that some aircraft were obtained in October 1920 but, as the country was occupied by the USSR in November, it is highly unlikely that they were ever used, let alone carried national markings.

Armenia became independent again in 1991 and was immediately in conflict with Azerbaijan. Ex-Soviet aircraft were used, still bearing red stars. The original roundel seems to have been in three colours, red, blue and yellow. This was soon followed by the addition of a white central disc.

The present marking consists of a roundel of the national colours, which have been noted in different sequences, but the usually accepted one is red (outer), then blue, orange and white.

An Armenian Su-25 Frogfoot, carrying one of several variations of the roundel.

AUSTRALIA

1942-46

1956

1982

The Australian Flying Corps was established in January 1913 and saw its first action in New Guinea in 1914. By 1915 a contingent of the corps was in action in Mesopotamia. The Australian Air Corps of 1920 became the Australian Air Force in March 1921, and the Royal Australian Air Force in June 1921. During this time standard RAF markings were used.

In June 1942 it was decided to paint out the centre red markings to avoid confusion with Japanese insignia, and this situation continued until the end of the Pacific war. In 1947 the centre red of the British roundel was, on the fuselage only, replaced with a red kangaroo. This marking was also used on the wings following a regulation of 26 September 1956. British rudder stripes or fin flashes were used throughout the history of the RAAF except for the 1942-45 period, when the red was eliminated.

From the late 1980s many aircraft used a low-visibility greyed-out version, or even a plain black or dark grey kangaroo.

Aircraft of the Royal Australian Navy from 1948 and the Army Air Corps from 1 July 1968 conformed to normal Australian use.

Above: *A Royal Australian Air Force British Aerospace Hawk with greyed-out low-visibility insignia.* Andreas Zeitler

A Pilatus PC-9.

AUSTRIA

The wing marking of an Imperial Austro-Hungarian Air Corps Albatros DV in the Military Museum, Vienna.
Author

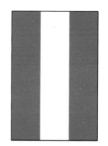

1914-15

1915-18

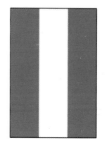

1918-20

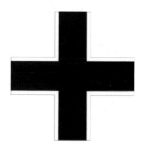

1918

1936-38, 1955

The Imperial Austro-Hungarian Air Service was officially formed on 9 August 1914, and aircraft were marked with the national colours of Austria – red/white/red – across the wing tips and on the rudder. By 1915 the German Balkan cross was used in addition to these stripes, and was soon adopted as the only marking on all but naval aircraft. The only differences between German and Austrian aircraft were that the former seldom used the white border to the cross, and had a different system of serial numbers. Austro-Hungarian naval aircraft continued to use the red and white stripes as well as the cross insignia and, on the white portion of the rudder, the coat of arms of Austria. Those in the naval service often had the arms of Hungary as well as those of Austria.

During the period after the First World War Austrian aircraft were in action against the fledgling Yugoslav air arm over the Carinthia region. Some aircraft reverted to the red and white stripes and some bore a thin black saltire cross across the fin and rudder.

Austria did not possess an air arm between 1919 and its reformation in 1936, when the marking chosen was a white triangle on a red disc, and rudders were marked red, white and red.

The paramilitary police used a roundel split horizontally in the national colours.

Austria was absorbed into Germany in 1938 and did not form an air force until 1955. The triangle in the roundel was revived, but without rudder markings.

AZERBAIJAN

This small Caucasian nation was independent from 1918 until April 1920. In March 1920 a number of aircraft were obtained from General Denikin's White Army and may have carried the general's blue triangle marking; it is very unlikely that specific Azerbaijani markings were used.

Azerbaijani MiG 29. via Greg Kozak

Independent again from 1991, ex-Soviet aircraft bearing red stars have been used in combat with Armenian forces. The first national marking was a roundel of blue, red and green with a crescent moon and white star in the centre. Azerbaijan's military aircraft now use a blue, red and green roundel split horizontally with the star and crescent superimposed on the central red.

BAHAMAS

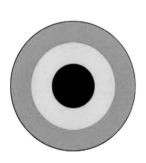

Bahamas Defence Force Vulcanair P.68. This Italian company produces mostly twin-engine aircraft for export. Andrew Martin

This ex-British colony became independent in 1973 and soon formed a small defence force. The air component consisted of one or two aircraft that carried the Bahamas Defence Force flag on their fin. Since 2009 a roundel of the national colours – blue (outer), yellow and black – has been in use.

BAHRAIN

Bahrain formed a Police Air Wing in 1965, and its helicopters carried the red and white Bahraini flag. In 1977 the Bahraini Amiri Defence Force was formed; its helicopters carried the national flag, while fixed-wing aircraft used the flag as a fin flash and, from 1985, a roundel format as a wing and fuselage marking.

Bahrain General Dynamics F-16. Andreas Zeitler

BANGLADESH

Formed in 1971, the air force uses an insignia based on the national flag. The fuselage and wing markings consist of a green and red roundel, while the fin marking is the national flag – a green field representing Islam and the country's agriculture, charged with a red disc for the blood of the martyrs.

Lockheed C-130. team_sjst

BARBADOS

The Barbados Defence Force was established in 1978 and acquired its first aircraft in 1981. This carries a civil registration and the national flag on its fin.

BELARUS

Once known as Byelorussia, this country has had a seat at the United Nations since 1945. It became fully independent in 1991 as Belarus and its air force continues to use the Soviet-style red star. A small version of the original red and white national flag was occasionally also used. From 2008 the new red and green national flag has been used as a fin flash, usually in 'wavy' format.

A Sukhoi Su-27 in Belarus. Like most of the country's aircraft, it bears the national flag on the tail and retains the red star on the wings. via Greg Kozak

BELGIUM

Belgium's first military aircraft were formed into the 'Company of Aviators' on 16 April 1913. In late 1914 the country's aircraft followed other Allied systems by adopting a roundel and rudder striping of the national colours – red, yellow and black were the ancient colours of the Duchy of Brabant, the basis of the modern Belgian flag.

    As the war clouds gathered in the late 1930s aircraft were

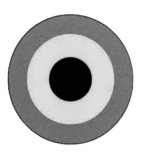

1915-40

1945-48

Navy

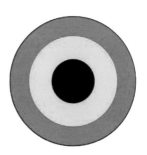

A Belgian roundel.
Daniel Brackx

camouflaged and the rudder striping was discontinued. After the German occupation in 1940 those aircrew who had escaped formed two squadrons in the Royal Air Force, Nos 349 and 350. They used RAF markings together a 'wavy' version of the Belgian flag below the cockpit. Normal Belgian markings continued to be used in the Congo.

On liberation in 1945 the roundel and fin flash were brought back into use. To follow Allied practice in the years after the war, the middle yellow ring was much thinner relative to the red and black rings.

Standard markings are now in use, the trend for low visibility accomplished by simply reducing the size.

Since the mid-1960s naval aircraft have featured a white anchor over the roundel.

Markings on a Belgian Navy Westland Wessex, 1965. Daniel Brackx

BELIZE

A Belize Britten-Norman Islander. BN Historians

British Honduras became independent as Belize in 1981, and the Defence Force acquired an air wing in 1983. Markings consist of the national flag, which is a horizontal tricolour of red, blue and red with the coat of arms on the blue centre.

BENIN

This former French colony became independence as Dahomey in 1960. The Dahomey Air Force, founded in 1962, used roundel and fin flash in the standard pan-African colours of red, yellow and green. The roundel was split vertically, the right half green and the left half split horizontally, yellow over red.

Benin Rockwell Commander 500 in 1991, having reverted to original Dahomey Air Force markings.

1962-75

On 30 November 1975 the country changed its name to the People's Republic of Benin, and its aircraft roundel became a green disc with a small red star, in line with the new national flag. By 1990 Benin had lost its Marxist government and replaced its national flag, and eventually its aircraft markings, with the original design.

1975-91

BHUTAN

This small Himalayan kingdom's Royal Bhutan Army has acquired a number of helicopters. Some reports have indicated the use of a yellow and orange roundel split diagonally, with a white centre. There is no confirmation of this and it may be a uniform patch badge. Any markings at all would probably be the national flag.

BOLIVIA

Bolivian Lockheed T-33.
George Trussell

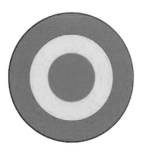

The Bolivian Aviation Corps was founded in August 1924 and changed its name to the Bolivian Air Force in 1940. Aircraft have always used roundels and fin or rudder markings in the national colours of red, yellow and green – red for the military valour of the people, yellow for mineral resources, and green for agriculture. In 1933, during the Gran Chaco war, some aircraft carried stripes in these colours instead of roundels; this was presumably to differentiate them from the roundels of the enemy, Paraguay.

BOSNIA-HERZEGOVINA

Srpska Soko Orao, with the later version of the roundel. The aircraft in the background carries a Serbian Air Force roundel. via Greg Kozak

This Balkan country has had a turbulent history. Governed by Turkey, Austria and Yugoslavia, and occupied by Italy and Germany, it only achieved a measure of independence in 1992. The few aircraft of the Bosnian Bosnia-Herzegovina Army carried the Bosnian national emblem of a blue shield with a diagonal white stripe and six fleurs-de-lys. The immediate effect was a virtual civil war.

The Bosnian Serb Republic (Republika Srpska) was formed in April 1992. Three main military aviation organisations were gradually formed – the militia, the army and the air force – and all used the Serbian flag of red over blue over white as a fin flash, but the wing/fuselage markings were different. Militia aircraft bore a red shield quartered with a white cross and a white border; in each quarter was an 'S'. The army used a blue shield with a yellow cross and yellow 'Ss'. The 'Ss' (actually the Cyrillic character 'C') represented 'Samo Slogo Srbino Sposavo' ('Only unity saves the Serbs'). From May to November 1992 the air force used a roundel split horizontally, red over blue over white. From November some aircraft used a normal roundel of red, blue and white in the centre.

The Bosnian Croat Federation, a Muslim and Croat joint force opposing the Serbs, was formed in the late 1990s, and from 1996 its aircraft marking was the Federation flag, a vertical tricolour of red, white and green; the white area bore a shield with the emblems of Bosnia, Croatia and the ten regions of the Federation area.

Bosnia 1992

Bosnia Croat Federation

Bosnian UTVA 75. Current Bosnian aircraft use just the national flag. via Greg Kozak

Bosnia Srpska

Serbian Army

Serbian Militia 1992

Bosnia-Herzegovina flag

The Dayton Accord of 1995 recognised the state of Bosnia divided between Bosnia Srpska and the Federation, then in 2006 the two military factions joined together. The initial marking reverted to the original Bosnian shield, but from 2007 aircraft were marked with the current Bosnian flag.

Bosnia Srpska

BOTSWANA

The Botswana Defence Force was formed in 1977, and aircraft are marked with the national colours of blue (for the sky), and black and white for racial unity. The insignia on the wings and fuselage take the form of a triangle. A thin stripe of blue over white over black is painted across the fin and the rudder

Botswana Britten-Norman Islander. BN Historians

BRAZIL

The Brazilian Navy formed an air service in 1914, and by 1918, when the Brazilian Army Air Service was formed, aircraft used a roundel and rudder striping of the national colours of green, yellow and blue. Army aircraft appear to have used rudder striping in just green and yellow. In 1934 the Army Service devised a different marking, consisting of a star with each of the five points split green and yellow. Superimposed on the centre was a blue disc with a white border that varied considerably in its width. The navy retained the original roundel and often added an adjacent black anchor. The

Low-visibility markings on a Dassault Mirage 2000. Chris Lofting

unified Brazilian Air Force was formed in 1941. Brazil declared war on the Axis powers and by 1944 had sent an Expeditionary Force to Italy. This used United States aircraft as supplied, but with the white star repainted with the Brazilian star. A design of the constellation of the Southern Cross was often added to the blue disc.

In 1961 the Navy Air Arm was again independent and reverted to the three-colour roundel.

Recently the Army Air Force changed its insignia to a three-colour roundel complete with the Southern Cross design and superimposed upon it a white sword.

Grey-out low-visibility markings are now often in use.

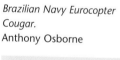

Brazilian Navy Eurocopter Cougar.
Anthony Osborne

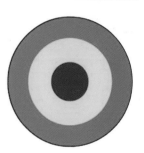

1914-18

1918-40

1943-45

Army

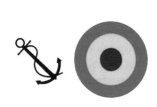

Navy

BRUNEI

An air wing of the Royal Brunei Malay Regiment was formed in 1965 at the time of the confrontation with Indonesia. The coat of arms of the regiment was carried with a fin flash of the national colours – yellow, white, black and red. A roundel version of the same colours with the addition of pan-Muslim green dates from about 1975 to around 1991.

Current Brunei aircraft carry the state arms outlined on a white disc if against a dark colour. Recent combat aircraft have appeared to revert to the four-colour roundel with national flag fin flash, but this was on Lockheed Martin promotion models in 2009 – no other evidence is yet forthcoming.

1965

1975-91

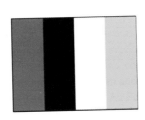

1975

BULGARIA

Bulgarian aircraft first saw action during the Balkan Wars of 1912 and 1913; their wing tips were painted red on one side and green on the other, and the rudders were marked with the national colours of white over green over red.

Bulgarian military aviation then ceased to exist until the country joined the Central Powers in 1915, and Germany supplied aircraft and most of the personnel. The aircraft bore normal German markings, mostly a black cross on a white square and white rudder. There were some reports of the addition of a green stripe along the trailing edges of the wings, but there is no official or photographic confirmation. During the summer of 1918 a black

1915-18

DFW B.1, 1918. Dimitar Nedialkov

Heinkel He 51; beyond, the twin-engined aircraft are Dornier Do 11s.
Dimitar Nedialkov

saltire cross on a white square was applied to the wings and fuselage of some aircraft. The fuselage, wings and fins of naval aircraft of 1918 carried special triangular markings of the national colours with a black lion rampant in the centre. White (outer), green and red roundels were used on Bulgarian aircraft to indicate those marked for destruction by the victorious allies.

After the war military aviation was forbidden and was moribund until the late 1930s owing to the economic state of the country. A small air force was formed in 1937, and after an initial use of a three-colour roundel an insignia was used based on the royal coat of arms, a rampant red lion on a red and yellow cross.

When the country joined the Axis powers in 1941, an insignia similar to that of the saltire cross of 1918 was used, and rudders were usually marked with the national colours. Bulgaria remained neutral in the Russo-German conflict but was forced, by the Russians, to change sides in 1944. Surrender markings consisted of a white stripe across the wings inboard of the black crosses and around the fuselage, and the cross marking was changed to a red and white roundel with a green horizontal bar, or, very occasionally, a white, green and red roundel.

A communist government took over in 1946 and this resulted eventually in the change of markings to the standard red star, with national colours as a small central roundel. Since the fall of communism in 1992 Bulgaria's military aircraft have carried a roundel of red and green with a white centre; naval aircraft carry, in addition, a black anchor and lion marking.

1918-20

1938-41

1941-44

1941

1944-46

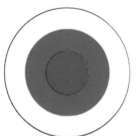

1946-48

Bulgarian-made Kazan KB-11 of 1937. Dimitar Nedialkov

LAZ-7, 1948. Dimitar Nedialkov

1948-92

Ilyushin IL-14, 1955. Dimitar Nedialkov

Navy 1918-20

Navy

BURKINA FASO

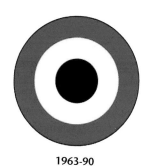

1963-90

This former French colony gained its independence in 1960, as Upper Volta. The Upper Volta Air Force was formed in 1964 and changed its name, with that of the country, to the Air Force of Burkina Faso in 1984. Before that year the insignia was a roundel of the previous national colours of red, white and black, but has since been changed to the pan-African colours of red, green and yellow. The current roundel is red over green, split horizontally, with a yellow star.

BURUNDI

This small African country became independent in 1962 and formed an air arm in 1966. The national colours of red, white and green are carried as a roundel, while some helicopters have been reported with the national flag.

CAMBODIA

Part of the old colony of French Indo-China, Cambodia became independent within the French Union in 1949, and fully independent in 1955. The Royal Khmer Aviation was founded in 1954, and the first markings consisted of the three towers of the temple at Angkor, in white on a red disc with a blue border. Fin markings, when applied, where blue over red over blue, with the temple illustration on the red.

T-28, 1970-75. Note the similarity to US markings, as Cambodia assisted with US forces in Vietnam.
via Kevin Curtis

MiG 21, 1989-94. via Kevin Curtis

1954-70

In 1970, when the United States began to help Cambodia as part of its involvement in the war in Vietnam, a new insignia was designed, a blue disc with the top left quarter in red; the red area carried a presentation of the three temple towers in white. Three small white stars also appeared on the top right of the disc, and on either side of the main disc were two bars in US Air Force style, i.e. white with a blue border and red central stripe.

When the Khmer Rouge took over the country in 1975, markings were changed again to a red star with a yellow temple. The invasion by Vietnam in 1979 saw the setting up of the People's Republic of Kampuchea. Vietnamese-supplied aircraft carried a very similar marking to the donor country, a yellow-bordered red disc and bars (the fin marking omitted the bars); a yellow temple replaced the Vietnamese star.

The Vietnamese withdrew in 1989 and the country's name became Cambodia once again. The marking then became a red-over-blue disc bordered in yellow with a central yellow temple. Since 1994 a return has been made to the original pre-1970 markings.

1970-75

1975-79

1979-89

1989-94

CAMEROON

Since its formation in 1960 the Cameroon Air Force has used a roundel of the pan-African colours. The rudder or fin flash originally carried two yellow stars on the green to represent the British and French Cameroons, but they were unified in 1975 and now use only one star.

Tail, 1960-72

Cameroon Air Force Aerospatiale CM170 Magister.

1960-72

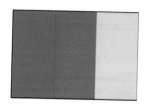

Tail

1946-47

CANADA

A Canadian Aviation Corps was in operation between July 1914 and February 1915, and aircraft tended to bear the insignia of the local flying school. The Canadian Air Force was established in 1920 and became the Royal Canadian Air Force in 1924. From 1920 normal Royal Air Force insignia was used, and although a wing and fuselage marking, replacing the centre spot with a maple leaf, was approved in 1940 it was rarely used. Aircraft engaged in active service during the Second World War bore the large squadron codes, but underlined in white; this system was abandoned in 1942. The occasional use of a green maple leaf superimposed on the red centre was never approved. Aircraft that may have come into contact with hostile Japanese forces in the Pacific had the central red spot painted out.

The red maple leaf centre was officially used from 19 January 1946. Initially it was a red leaf on a solid blue disc, but British-style markings were soon adopted, the maple leaf replacing the centre red spot. The navy tended to use a wider blue outer ring, the air force the more normal type. The fins bore British fin flashes, with the thin white band, until 1958.

Between 1958 and 1965 the then current Canadian flag was used on the fin. The Canadian flag changed in 1965, and the new one was now used as a fin flash. Low-visibility marks in pale grey or black are in frequent use.

1947-58

Tail, 1958-65

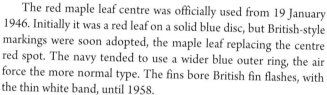

Low-visibility markings on a C-130.

1958-65

Tail, 1965-88

1965-88

1988

CAPE VERDE

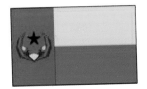

Tail, 1976-92

Tail, 1992

The few aircraft of this small former Portuguese colony used the national flag in the pan-African colours between 1976 and 1992. A change of flag was then reflected in the aircraft markings.

CENTRAL AFRICAN REPUBLIC

Central African Republic Britten-Norman Islander.

This former French colony, briefly called the Central African Empire, formed a small air arm in 1960. The roundel and fin flash are based on the national flag and consist of the pan-African colours and blue and white to symbolise unity with the former colonial power.

CHAD

Another former French colony, Chad formed its air force in 1960 with gifts of French aircraft. It has always used a roundel and fin flash or rudder striping in the national colours of red, yellow and blue. In 2008 the roundel was changed to a three-colour segmented format.

1960

Chad Pilatus PC-7 with original markings.
via Greg Kozak

Sukhoi Su-25 in the country's current markings.
via Greg Kozak

CHILE

The Chilean Military Aviation Service was established in 1913, and became the Chilean Army Aviation in 1918. The Naval Air Service was formed in 1921 and these air arms were amalgamated to form the Chilean Air Force in 1930.

1918-30

Chilean Air Force Bell 412.
Anthony Osborne

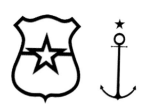

Navy, 1983

Navy

Chilean Navy Embraer EMB110 Bandeirante. via Greg Kozak

Up to about 1920 aircraft were supplied by Great Britain, and as red, white and blue were the Chilean national colours the British markings were retained. In 1921 a new insignia was adopted consisting of a shield split horizontally into blue over red with a white star in the centre. Rudders were painted dark blue and charged with a white star, but some naval aircraft bore a white anchor instead of the star. From 1929 to about 1930 aircraft used a roundel of red (outer), white and blue with a white star on the blue. Most aircraft continued to display a blue rudder and white star although there was occasional use of three-colour rudder striping with red forward. In 1930 the shield emblem was chosen as the insignia of the new Chilean Air Force. There have been reports of a red and blue roundel with a white star, but these are not confirmed.

The Chilean Army formed its own aviation unit in 1970, and the same insignia was used but the rudder was red instead of blue. The navy air arm has used a black anchor on a blue rudder since its formation in 1954. Some aircraft now use a low-visibility version consisting of a black outline of the insignia.

CHINA

The history of the markings of Chinese military aircraft can be very confusing. Central government was only achieved in 1927, but much of the country was controlled by up to thirty warlords, most of them with their own air forces.

The first Chinese military aircraft were based at the flying school near Peking in 1913, and these carried a star on the wings, fuselage and fins. Each of the five points of the star was a different colour, representing the main ethnic groups in the country: red for Manchus, yellow for Han Chinese, blue for Mongols, white for Muslims, and black for Tibetans. There was also some later use of roundels and fin flags in these colours.

1913-16

Tail, 1928-30

Curtiss P40 of the American Volunteer Group, helping the Chinese against Japan in 1942.

Chinese civilian C-47 being used as a military transport towards the end of the war. George Trussell

1928-30

1936-49

Sun Yat-sen's Kuomintang government set up its own air force at Nanking in 1923. The aircraft used the twelve-pointed white star on a blue disc, still used in basic form in Taiwan. Between 1923 and 1927 the roundel normally included an outer red ring. In 1928 the Chinese central government under Chiang Kai-shek formed its own air force, which also used the twelve-pointed star (the points represented Chinese time – the day divided into two-hour periods and the year into twelve months). There were various rudder markings, but by the late 1930s these were fixed on either blue and white horizontal stripes or the national flag of red with a canton of blue bearing the white star as a fin flash. The flag marking was dropped in about 1939.

The American Volunteer Group, which flew for the Chinese against the invading Japanese, dispensed with any fin markings, but the Chinese Air Force continued to use this marking until defeated by the Communist People's Army in 1949, when the government fled to Formosa (Taiwan).

The Chinese National Aviation Company was founded as a civilian airline in 1929. During the war with Japan its aircraft were camouflaged and carried a white Chinese ideogram, probably simply meaning 'China', on a blue disc. Although civilian, the aircraft were used in many clandestine activities between 1937 and 1945.

1946-49

Tail, 1946-50

Ex-Japanese Kawasaki Ki-48 in the early People's Army, 1945. George Trussell

1946-50

National Government of
China (on Japanese side)

Manchuria, 1928-31

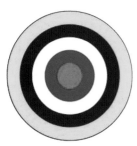

Manchuria, 1931-45

Manchuria, 1942-45

Nanchang CJ-6 showing current markings. George Trussell

PEOPLE'S REPUBLIC OF CHINA

The attempt by communist forces to seize power in China began in the late 1920s. The first aircraft marking, dated 1930, was a simple red star with the name 'Lenin' on the fuselage in red. Sinkiang in western China was taken over by Mao Tse-tung's force in the 1930s and an aviation school was established with Soviet support. The fuselage of its aircraft was marked with a stylised comet – a six-pointed star with a white 'tail' – although some retained their original blue and white roundel. By 1944 the Red Army had captured a number of Japanese aircraft, and the Japanese roundels were covered with a Chinese ideogram representing the forces.

Between 1945 and 1950 there were many different variations of red stars, some bearing a Chinese marking representing 1 August 1928, the date of the founding of the Red Army. After 1950 a red star with the ideogram in yellow was flanked by yellow-bordered red bars. Since 2010 a number of aircraft have been seen without the bars.

The Chinese Navy established an air arm at Foochow in 1916. From 1927 a black anchor was added to the roundel, and in 2007 yellow stripes were added to the red side bars.

Illustrations of a selection of warlord markings and those of pro-Japanese organisations of the Reformed Government of China and the National Government of China (erroneously referred to as the Cochin China Air Force) are also shown here.

MANCHURIA

The warlord Chang Tso-lin ruled Manchuria up to 1928, when Chang Hseuh-tang succeeded him. In the 1920s they survived with a motley collection of First World War aircraft, which by 1930 carried markings in the form of a blue disc with a small white twelve-pointed star. The rudder was split blue over red with the star

Navy

on the blue. Japanese forces occupied Manchuria in 1931, renaming it Manchukuo. A national air force, the MKKK, was established, but the aircraft carried military markings of a roundel in yellow, black, white, blue and red. The tip of the fin was similarly marked. The Manchukuo Air Force was formed in 1942 using Japanese aircraft, which carried roundels of a yellow disc tipped with the other four colours. The force was disbanded in 1945.

Hong Kong and Taiwan are discussed separately.

COLOMBIA

The Colombian Air Force was founded in 1922, and the initial roundel was red (outer), yellow and blue, or yellow (outer), red and blue. This was later changed to a roundel of intricate design based on the same proportions as the national flag, in the centre of which was superimposed a white nine-pointed star. The current white five-pointed star dates from about the 1950s, and naval aircraft mark the standard roundel with a black anchor. Greyed-out low-visibility markings are now used on some aircraft.

1925-53

Navy

COMOROS

This group of islands in the Indian Ocean became independent from France in 1975. The very few government aircraft were marked with the national flag of green with four white stars for each island. The marking was changed with the national flag in 1992.

CONGO

This Central African country has become known as Congo-Brazzaville to distinguish it from its larger neighbour. It gained its independence from France in 1960 and its small air arm used only a fin marking of the national flag, red and green split by a yellow diagonal stripe. The country became a People's Republic in 1970 and the marking was changed to a red disc with a green wreath of leaves, a crossed yellow hammer and hoe, and a yellow star. 1990

Congo 1970-90

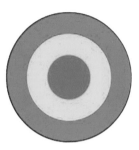

Congo-Brazzaville

Democratic Republic 1997-2005

saw a change back to the Congo Republic, using the original colours in a roundel format of red, yellow and green. The centre spot is seen as black owing to the shortage of green paint.

CONGO, DEMOCRATIC REPUBLIC

This large, central African state gained its independence from Belgium in 1960 as the Congolese Republic. The new national flag was blue, split diagonally with a yellow-bordered red stripe; a yellow star on the upper part of the flag was added later, which was also the fin flash. The roundel appears to be a yellow star on a blue disc. In 1964 yellow-bordered red bars were added to the disc.

In 1972 the country changed its name to Zaire, and the new flag was green with a yellow disc bearing an arm carrying a torch. This was used as a fin flash with a roundel form for wings and fuselage.

In 1997 Zaire became the Democratic Republic of Congo with a new flag, blue with a large yellow star and six smaller stars. This was used as a fin flash until the flag reverted to the pre-1972 version in 2005. The marking has also changed back, but the side bars are orange with a red border surrounding the entire roundel.

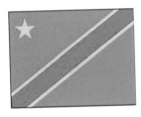

Democratic Republic, tail, 2005

KATANGA

This breakaway province formed its own air force during its brief period of independence between 1960 and 1962. The fin marking was the national flag of red over white split diagonally with a green stripe; on the white area were three orange crosses. Roundels were red, green and white with the crosses on the white centre spot. Some later aircraft may have carried a small single cross under the cockpit, but most carried no markings at all.

Democratic Republic 2005

Congolese Republic 1960-64

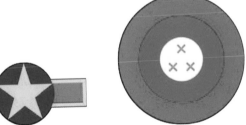

Congolese Republic 1964-72

Katanga

Zaire, 1972-97

Left: *An Aermacchi MB326 bearing the current flag of the Democratic Republic of Congo.*

Right: *A Congo-Brazzaville Nord Noratlas carrying the current roundel.*

A Congo Democratic Republic Mil-8. Guido Potters

COSTA RICA

Although it could be said that Costa Rican military aviation began in 1928, it was 1948 before any real attempt was made to form an air force. This force lasted barely a year and was marked with an insignia identical with the pre-war US insignia but with the colours reversed. The air force was revived in 1955 and aircraft were marked with a variant of the national flag. The rudder was blue, white, red, white and blue, and an elongated version of the flag was used on the wings and fuselage. Since the 1970s aircraft have used a roundel almost identical to the British one, but on the outer ring is the inscription 'Ministerio de Seguridad Publica Seccion Aerea'.

Aircraft now carry the national flag alone.

1948-49

1955

1941-44

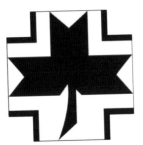

1944-45

1991-94

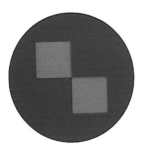

1994

CROATIA

With the Italian and German attack on Yugoslavia in 1941, the country was split up and Croatia became independent as an Axis supporter. The Croatian Air Force used the ancient red and white chequered shield of Croatia on wings and fin and sometimes on the fuselage. Aircraft for the Russian front and the German-commanded Croat Legion used standard German insignia and the legion badge under the cockpit. From 1944 aircraft carried the black cross of King Zvonomir, an ancient ruler of Croatia, on the wings and fuselage. The fin carried the red and white shield surmounted by a framed letter 'U', representing the Ustachi – the Axis-controlled government.

Croatian Avia F13 of 1945. Zoltan Rados

A Mil-8 of 1992 with the low-visibility shield marking. Zoltan Rados

Mil-24 with the 1992 shield marking. Zoltan Rados

A Zlin 242 carrying current markings. Zoltan Rados

Croatia was absorbed back into Yugoslavia in 1945, but became independent again in 1992. The new Croat Air Force used the same chequered shield, now surmounted by five small shields representing Ancient Croatia, Dubrovnik, Dalmatia, Istria and Slavonia. In 1993 a much simpler marking was introduced – a blue disc charged with two red squares. The Croat national flag acted as a fin flash.

KRAJINA

In 1993 Serb forces invaded southern Croatia and set up the Republic of Krajina Serbia. Military aircraft were marked with the Serbian flag as a fin flash. A semicircle of red over blue over white was inscribed with 'Krajina' in red and 'Militia' in white. Croat forces reoccupied Krajina in 1995.

CUBA

Cuban military aviation can be traced back as far as 1913, but it was after the First World War that an organised air arm was formed. These first aircraft bore the current United States markings; all the aircraft came from there, and Cuba shared the same national colours. From 1928 a new insignia was used, a blue disc charged with a red triangle, point down, with a white star in the centre; the rudder carried a version of the national flag.

1928-55

A new marking was introduced in about 1955. This followed United States usage by having a white star with a red border and blue bars on each side. The rudder markings were retained.

From 1959 into the 1960s Castro's Revolutionary Air Force adopted yet another type of insignia. This consisted of a red triangle with a white star and bars on each side of blue, white and blue. This marking was also used by the clandestine 'Liberation Air Force', which took part in the ill-fated 'Bay of Pigs' episode.

1955-59

During the early 1960s the Cuban Revolutionary Air Force reverted to the pre-1955 markings. Some of the Soviet aircraft used by Cuba appeared to carry wing and fuselage markings of a plain red triangle and white star.

The Cuban Navy obtained some aircraft as early as 1928, and during the 1930s some of them carried two black crossed anchors on the fin.

1959-62

CYPRUS

The Cyprus National Guard was formed in the late 1960s and its aircraft bore the national flag consisting of a map of the island in orange with a green wreath of leaves. As this force only operates in the Greek half of the island, aircraft now use Greek blue and white roundels; recently a few of these had the white area thinner than the blue. The flag is still used as a fin flash.

Aerospatiale Gazelle operating in the Greek part of Cyprus.

1969

CZECH REPUBLIC

Czechoslovakia became an independent state on the dissolution of the Austro-Hungarian Empire in 1918. The national colours chosen were a combination of the red and white of Bohemia and the blue and white of Slovakia. The first insignia used the national colours as a roundel of red, blue and white, or in reverse, which was identical to that of the Imperial Russian markings; as most of the new Czech Air Force's aircraft came from Russian sources, this was convenient. By 1920 the national flag had been officially adopted, the white and red of Bohemia with a triangle of blue for Slovakia, and this was marked across the wings and fin. From 1926 there was a roundel version.

In 1938 German forces invaded Czechoslovakia and the country ceased to exist. It was split into the Protectorate of Bohemia-Moravia and the republic of Slovakia, which had a desire to become a separate state. Slovakia formed its own air force in 1939. Many Czech airmen fought on the side of the Allies during the Second World War; Royal Air Force squadrons 310, 311, 312, and 313 were all Czech units and all their aircraft carried a small Czech roundel in addition to the British markings. A Czech Legion was formed in Russia, which from late 1944 used the pre-1938 roundel.

1918-20

1920-21

1921-39

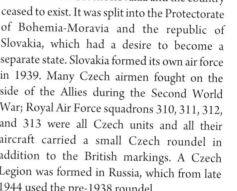

Left: *Low-visibility markings on a Saab Gripen.* Jozef Grego

Below: *Czech Republic MiG 21 showing the current markings as also used by Czechoslovakia from 1921 to 1993.* Jan Jorgenson

After liberation Czechoslovakia was reunited and began to build up its air force. Various markings were used, including the old flag type, but by 1948 the roundel form was in normal use. The white segment was always outboard and was surrounded by a blue or white ring on dark surfaces. Unusually for Warsaw Pact aircraft, the roundel also appeared on upper wing surfaces.

From the 1930s a paramilitary organisation called the Air Police operated. This used civil registrations, but with an oblate form of the roundel.

In 1993 the country decided to split into the Czech Republic and Slovakia, and the Czech insignia continued to use the same markings. Slovakia is described separately.

DENMARK

1916-40

The Danish Army and Navy operated separate air arms from 1912 until the formation of the Royal Danish Air Force in 1950. The red flag and white cross are reputed to date from 1219, and a swallowtail version was used on the rudders of all naval aircraft. Army aircraft used the simple red and white roundel on the wings, fuselage and, initially, rudder.

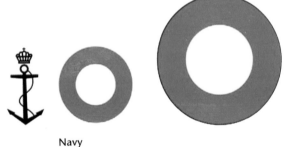

Navy

Denmark was occupied by German forces between 1940 and 1945, but Danish-marked aircraft were used in Greenland and for training at Allied bases. Since 1950 all aircraft have used the roundel and swallowtail. Naval aircraft often use a crown, rope and anchor in addition to the Danish markings. The low-visibility trend of the 1980s saw a reduction in the size of the insignia.

Danish Westland Lynx.
Jan Jorgenson

DJIBOUTI

This small African country was previously known as French Somaliland, then as the French Territory of the Afars and Issas. An air arm was formed in 1979 but defence and policing has been dealt with by French forces based in the country. The national flag, which is used as a fin flash, contains green for the Afars and blue for the Issas. The first insignia was a roundel split into three segments, blue, green and white, with a red star on the white. The roundel now in use follows normal practice of concentric colours with a red star on the central white spot.

1982

Djibouti Nord Noratlas.
via Kevin Curtis

DOMINICAN REPUBLIC

The Dominican Republic occupies the eastern two-thirds of the island of Hispaniola in the West Indies. The first attempt at forming an air arm took place in 1930, and the first aircraft were painted with a red, white and blue cheatline along the fuselage. From 1933 the rudders of the Dominican aircraft were marked with the national flag of a white cross on a red and blue field, and a roundel form was used on the wings. This marking has remained unchanged except for a reduction in size for low visibility.

Dominican Republic CASA 212.
via Greg Kozak

Navy

ECUADOR

Since its formation in 1920 the air force of Ecuador has always used a roundel and rudder striping or fin flash based closely on the country's national flag, which is a horizontal tricolour of yellow, blue and red, the yellow taking up half of the flag. These proportions are followed on the insignia. During the conflict with Peru in 1941 some aircraft bore stripes of the national colours below the wings.

The navy's air arm was formed in 1967, and markings consist of a black anchor superimposed on the roundel. A fuselage marking of a black anchor design on a white disc is often used on helicopters with the national colours on the rear boom.

Ecuador Dassault Mirage 5. Mario Overall

EGYPT

1932-53

A decision was taken in 1930 to form an Egyptian Air Force, and it became established in May 1932. The first aircraft wore green-painted wing tips marked with a white crescent moon and three stars, symbolising the three main religions of Egypt – Islam, Christianity and Judaism. A 'wavy' version of the Egyptian flag was carried on the fuselage and the rudder was striped green, white and green. Later, in 1932, a roundel form for wings and fuselage was adopted; this was green, white and green with the crescent and three stars on the central green. The force became the Royal Egyptian Air Force in 1939, whereupon a white crown was added to the outer ring of the insignia.

1932

Egyptian MiG 23. George Trussell

At the end of the Second World War some British-supplied aircraft were used, their roundels having the red and blue painted over with green and without the stars and crescent. With the fall of the monarchy in 1953 a return was made to the original 1932 markings without the white crown.

In 1958 Egypt formed, with Syria and Yemen, the United Arab Republic (UAR), which would have a joint military command and a similar flag of red over white over black. The white part of the flag, and the insignia, carried a number of green stars, in Egypt's case two. Until 1971 Egypt continue with the title UAR but dropped the two green stars and replaced them with the golden falcon, and in 1984 with the golden eagle of Saladin. These changes were reflected on the fin flash but since 1971 the wing and fuselage markings have been a plain red, white and black.

1945-53

1953-58

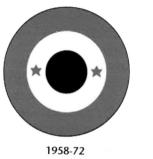

1958-72

1972

EL SALVADOR

From hesitant beginnings as early as 1917 the air force of El Salvador was officially formed in 1922. The wing and fuselage markings have always been a plain blue, white and blue roundel. Horizontal rudder striping was changed to vertical in about 1930. Since the 1960s the horizontal national flag has been used as a fin flash and fuselage marking.

1965

El Salvador Chance-Vought Corsair FG1 during the 1969 war with Honduras. via Kevin Curtis

Cessna A-37 in current markings. George Trussell

2007

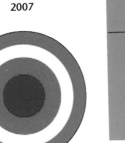

EQUATORIAL GUINEA

This small African country gained its independence from Spain in 1968. A few non-combatant aircraft were obtained in 1972 and bore only the national flag on the fin. Since 2006 aircraft have carried a roundel of green (outer), white, red and blue

Equatorial Guinea Mil-24.
Keith Parkinson

1991

ERITREA

After securing independence from Ethiopia, Eritrea formed an air force in 1994. Markings followed the national flag with a roundel version, segmented in green, red and blue. Initially a few aircraft had a yellow star on the red section, but since the late 1990s a yellow wreath has been used. The national flag has had occasional use as a fin flash instead of the roundel.

1992

ESTONIA

Estonia achieved independence from Russia after the revolution, and in 1918 an air force was quickly formed. The insignia consisted of the national colours of blue, black and white in triangular form. The point of the triangle faced backwards on the wings, and downwards, or sometimes backwards, on the fuselage. The rudder was striped horizontally in the national colours.

Estonian Aero L-39, carrying markings used during the country's previous period of independence in 1918-40. Jan Jorgenson

The country was absorbed into the USSR in 1940, but became independent again in 1990, and a return was made to the original markings.

ETHIOPIA

Ethiopia is one of Africa's earliest independent nations, and apart from the period between 1936 and 1941 when it was part of Italian East Africa it has remained so. The green, yellow and red colours date from 1897 and were the original pan-African colours.

An air arm was formed in 1924, and until 1936 bore the Ethiopian flag as a wing, fuselage and fin marking. The air force was reorganised in 1946 and aircraft were marked with a roundel of green, yellow and red, a yellow star design being marked on the central red spot; no rudder or fin markings were carried. This design has been changed several times. The original version was a stylised six-pointed star based on the Star of David (the Emperor of Ethiopia claimed the title of 'Lion of Judah'). This was followed, in 1974, by a more clearly designed six-pointed star. After the revolution the symbol was painted out and replaced in 1977 by a plain five-pointed star. The current roundel dates from about 1998 and is a complex marking featuring the name of the country and the words 'air force'. It also includes the national flag and a wreath of leaves.

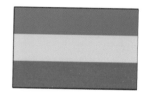

1924-36

1946-74

1974-84

1918-44

FINLAND

Finland became independent after the break-up of the Russian Empire. Blue and white have always been the Finns' national colours, so were chosen for the national flag and the air force's insignia. Count von Rosen of Sweden was very involved with the setting up of the first Finnish Air Force, so his personal swastika emblem was used as the insignia on wings, fuselage and, in the early days, the rudder.

Soviet forces invaded Finland in 1939, and following the German invasion of Russia in 1941 Finland threw in her lot with Germany. In 1944 the country was forced to sign a peace treaty with the Soviet Union and to declare war on Germany. A swastika marking was obviously inappropriate, so a white, blue and white roundel was adopted. In the 1980s low-visibility trends saw a considerable reduction in the size of the insignia.

Me-109G in Finland in 1943. via Kevin Curtis

Finnish Piper Chieftain. Jan Jorgenson

FRANCE

To France goes the honour of being the first country to officially implement a military aircraft insignia, in a decree of 26 July 1912. The colours of red, white and blue, used since the revolution, were used in a roundel on the wings, and the rudder was striped with these national colours. The use of the roundel on the fuselage has only been normal since 1918, and is still in current use.

Polish airmen who flew for France in the 1940 campaign were allowed to mark the Polish national insignia on the fuselage sides while retaining standard French markings on the wings and rudder. After the French defeat in 1940, the country's armed forces split. The Vichy French Air Force, which collaborated with the Germans, used normal insignia, but, owing to the possibility of confusion with the RAF roundel, aircraft were additionally marked with red and yellow stripes on the fin, cowlings and tail planes. Other features were a full-length white fuselage stripe, and the occasional use of thin red, white and blue under-wing stripes. The Free French Air Force used normal RAF or US markings, with in addition a red or blue Cross of Lorraine on a white disc. Later, French roundels were used and the Cross of Lorraine was used equal in size to those on the wings and sometimes replaced on the fuselage. In the late 1940s French aircraft followed Allied practice by marking roundels and fin flashes with a much reduced white area.

From the 1960s to the 1980s it was normal to outline roundels with a thin yellow border, but this has now been almost totally discontinued. Since about 1925 naval aircraft have used a black anchor on the roundel and rudder stripes or fin flash.

1912-40

1945-48

1948

French Dassault Mirage 2000. Jan Jorgenson

Free French 1940-45

Navy

GABON

This former French territory, independent since 1960, soon formed a small air arm. Aircraft were marked with a roundel and fin flash in the national colours of blue (outer), yellow and green. From 1970 to 1982 the colours were in reverse order, but this may have indicated aircraft of the Presidential Guard.

Gabon Aerospatiale Alouette III.
via Greg Kozak

GAMBIA

This small West African nation bears the national flag on the fin, consisting of red over blue over green, with a thin white stripe dividing the colours.

GEORGIA

This ancient state became independent from Russia in May 1918. Throughout 1919 and into 1920 there was considerable British involvement, including military assistance and the supply of aircraft; Italy also provided aircraft in late 1920. Some of these machines bore the 6th-century Georgian insignia, consisting of a cross, black or dark red, on a white disc on the fuselage sides and wings; there may also have been an added thin black border. Georgia was absorbed into the Soviet Union in February 1921.

The country became independent again in 1991 and used former Soviet aircraft with red stars. Very soon aircraft were marked with a dark red seven-pointed star on a white disc with a black border. After a change of flag in 2004 the colour of the star was changed to bright red.

1920

1991

2004

Below right: *Aero L-29, showing the new red star.* via Greg Kozak

Below: *Georgian Sukhoi Su-25 with the original dark maroon star.* via Greg Kozak

GERMANY

The German Army obtained its first aeroplanes in 1910 and the German Navy in 1912. Before the First World War national markings consisted of a black stripe across each wing. A roundel of the national colours was considered but never implemented, presumably because shape rather than colour was the important factor in identifying friend or foe. On 28 September 1914 the German Army adopted a black Maltese cross insignia, which was displayed on the wings and across the fin and rudder, outlined in white when against a darker colour. By 1915 the crosses were also applied on the fuselage. On 15 April 1918 the cross was changed to a straight-edged so-called 'Balkan' cross. Some Bavarian units adopted black and white striped fuselage markings, and an all-black fin and rudder.

Germany's defeat in November 1918 led to the complete destruction of its air force, but by the late 1920s various clandestine attempts were being made to form a new air arm. These aircraft carried no national markings. After 1933 aircraft, military in all but name, were being marked with rudder stripes in the national

1912-14

1914-18

1918

Second World War markings on a German Me 410.
Author

East German MiG 17. Yefim Gordon

1933-36

Tail, 1936-40

1936-40

Tail, 1949-45

colours of red, white and black on one side and a red band with a black swastika on a white disc on the other. The straight crosses on the wings and fuselage and the black swastika on the fin were the standard markings throughout the Second World War, although as the war progressed many variations were used.

In 1945 the country was divided into four zones of occupation. In 1949 the American, French and British zones became the Federal Republic of Germany (West Germany), and the Soviet zone the German Democratic Republic (East Germany). Both Germanys used the same flag, a horizontal tricolour of black, red and yellow.

A small air arm attached to the People's Police was formed in East Germany in 1950, and the East German Air Force in 1955. These aircraft bore the national colours in a diamond shape with a black border. In 1959 the state arms was added to the flag and to the red central portion of the insignia. This marking was abolished upon the unification of Germany in 1990.

The West German Air Force was established in 1955 and aircraft carried the Maltese-style cross, which had last been used in early 1918. The national flag was used as a fin flash.

Naval aircraft bear an anchor device in a circle forward on the fuselage. West German markings continued as the insignia of the German Air Force.

Panavia Tornado. Anthony Osborne

German Navy Westland Sea King. Anthony Osborne

1940-45

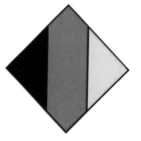

East Germany, 1955-59

West Germany, 1955

Navy

East Germany, 1959-90

GHANA

Formerly the British Gold Coast, Ghana was one of the first of the newly independent African states. In honour of the very first, Ethiopia, it chose that country's red, green and yellow as its national colours. With the formation of the Ghana Air Force in 1959, these colours were adopted as a roundel. The black star, which featured on the central yellow part of the flag, was used as a fin marking. In 1964, for political reasons, the yellow part was changed to white on the flag and the insignia, and in 1965 the national flag began to be used as a fin flash. In 1966 the white on the flag and roundel was changed back to yellow

Tail, 1959-64

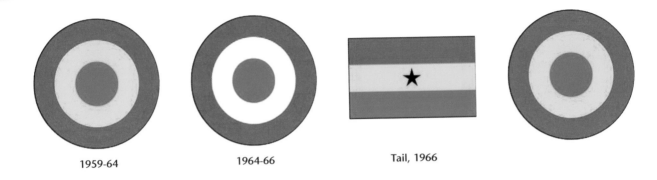

1959-64   1964-66   Tail, 1966

GREECE

Greek military aircraft took part in both the 1912 and 1913 Balkan wars. These machines carried blue and white stripes on the wing tips, front stabilisers and tail planes.

During the First World War separate air arms for the army and navy were formed. The Army Air Corps was founded in September 1917 under French command and therefore used French markings; the British-led Navy Air Corps aircraft carried British markings. On the return of these forces to Greek command in June 1919, blue, white and blue were used for the roundels and rudder striping. Until the formation of the Royal Hellenic Air Force in 1931 the army and navy retained separate air arms. The army insignia was a much lighter blue than that of the navy, and the navy also often used an anchor on the rudder. The pale blue marking was used for all aircraft from 1931.

1912-14

1919-31

1931

Greek Lockheed-General Dynamics F-16 Block 52. Greek Air Force

During the Second World War the Royal Air Force formed Greek units as Nos 335 and 336 Squadrons. These carried a Greek flag as well as the RAF markings and the spinners were often painted blue and white. On 1 February 1945 these squadron aircraft were passed to Greece, which changed the markings to the pre-war designs and used a blue and white fin flash.

GRENADA

This West Indian island became independent in 1974. In 1979 a Marxist government seized power, and in 1983 an air arm called the People's Revolutionary Government Air Wing was proposed. The island was invaded by US-led troops in that year and the air wing was never established.

GUATEMALA

The origins of military aviation activity in Guatemala can be traced back to 1914, but it was not until 1923 that the Guatemalan Air Force was formed. Markings took the form of a white five-pointed star with a blue centre spot, and rudder striping was vertical blue/white/blue, insignia that is still used today. In 1929 some French aircraft were supplied with erroneous blue/white/blue roundels. From about 1939 to 1947 a six-pointed star was used, presumably to distinguish Guatemalan aircraft from those of the United States. Blue, white and blue stripes have been noted under the wing tips of some aircraft.

1923-39

1939-47

Ryan STM. Mario Overall

GUINEA

Guinea formed its air arm in 1959, soon after independence from France. It has always used the pan-African colours as a fin flash, but roundels have only been in use since the 1980s.

GUINEA-BISSAU

This former Portuguese colony became independent in 1974 and formed an air force with Soviet assistance in 1978. Its insignia consists of the black star of Africa on a red disc.

Guinea-Bissau Mil-17. via Greg Kozak

GUYANA

Once known as British Guiana, this South American country became independent in 1966. The Guyana Defence Force Air Wing was formed in 1968, the Air Command in 1970. The insignia is a roundel of black, yellow and black, taken from the national flag.

HAITI

1943-71

The Aviation Corps of the Haitian Guarde was first formed in 1943. The first insignia was a disc, horizontally divided blue over red with a white border, and usually, but not always, with side bars, following US usage. In 1964 a return was made to the pre-1806 national colours of red and black, which was reflected in a change to the insignia. In 1971 the blue was reinstated and the insignia took the form of a blue and red roundel, with white side bars.

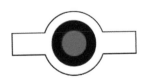

HONDURAS

Military aviation in Honduras dates from 1921 when the first of a succession of mercenary pilots and their aircraft were employed by the government. Aircraft were painted overall in the national colours of blue and white, usually combined with a blue/white/blue rudder and the national arms on the fuselage.

1936-75

This Honduras Cessna 185 carries current markings.
via Greg Kozak

The Honduran Air Force was officially formed on 25 February 1936, and with the advent of American aid during the Second World War a specific national insignia was formulated. This was a blue/white/blue fin and wing tips, the white area containing a blue star. With the arrival of jet aircraft in the 1970s the national insignia was relegated to a small Honduran flag on the fin.

HONG KONG

The original Hong Kong Volunteer Air Force was formed in the 1930s and saw activity, but not combat, with the Japanese forces in 1941. The defence force, with an air arm, was reformed in 1946 and became the Royal Hong Kong Auxiliary Air Force in 1970. Until this time standard RAF markings were used, but by the 1980s

1982-97

A Slingsby T67 Firefly of Hong Kong's Auxiliary Air Force.

Sikorsky S-70 Black Hawk. via Greg Kozak

several aircraft were marked with red wing tips and rudders, a red fuselage band edged in white and blue, and the Hong Kong coat of arms on the fin.

Although returned to Chinese control in 1997, Hong Kong has the status of a Special Administrative Region. Government aircraft bear a red and white roundel that includes a white five-petal lotus flower and inscriptions in Chinese reading 'Chinese Government Flying Service of Hong Kong SAR', and 'Hong Kong' in English. All aircraft carry Chinese civil registrations.

HUNGARY

1919

During the First World War Hungarian aircraft were marked as those of Austria. Some naval aircraft, however, unofficially carried the shield of Hungary as well as that of Austria on the fin and rudder.

After the collapse of the Austro-Hungarian Empire in November 1918, a Hungarian Republic was set up on 24 March 1919, riven by internal factions. Many of the aircraft left in the country were, officially, marked across the wings with a red, white and green chevron design, a 15th-century Hungarian symbol. Out of the chaos communists finally appeared as the strongest group and a Hungarian Soviet Republic was established in April 1919, which was immediately invaded by Czechoslovakia, Romania and Yugoslavia. To defend the country a Red Air Corps was organised with a large collection of First World War aircraft and veteran aviators; the corps used a Soviet red star as a marking, usually on a white panel. A white star on a red panel was occasionally used below the wings.

1919

On the collapse of the Soviet Republic in August 1919, the air corps was disbanded and military activity was forbidden in Hungary. However, a clandestine air force was set up in the late 1920s marked with civil registrations but often, in addition, various forms of the national colours. The air force was officially born on 23 August 1938, and became the Royal Hungarian Air Force on 1 January 1939. Aircraft were marked with a new red, white and green chevron design across the wings and fin.

1938-41

With the participation of Hungary in the Russian campaign in 1941, its national insignia was brought into line with the other Axis countries by adopting a cross-type marking. This was introduced in

1942-45

MiG 21 showing 1980 markings. Author

March 1942 and consisted of a black square bearing a cross in white, and later in varying shades of grey. Initially fins and elevators were striped in the national colours, but this was soon abandoned.

On 14 April 1948 a new air force was formed with a new marking, a green spot on a white triangle in a red disc. After the communist takeover this was changed on 15 November 1949 to a red star inside two thin rings of red and green.

On 16 June 1951 Hungary joined Bulgaria and Romania by adopting an insignia consisting of small roundel of the national colours superimposed on the centre of a red star. This was used until the collapse of the communist regime in 1990. On 31 January 1991 a return was made to a modified version of the chevron design.

During the uprising of October 1956, a Hungarian flag was painted over the red stars on rebel-held aircraft.

1944-45

Hungarian 'shark-mouth' L-39. Jan Kucera

1948-49

Saab Gripen with low-visibility markings. Anthony Osborne

1949-51

1951-91

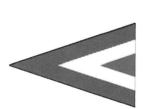

Insurgents, 1956

1991

ICELAND

Iceland has never possessed any armed forces. Government-owned aircraft engaged in fisheries protection and coast guard duties carry normal civil registrations and the national flag on the fin. Some have been noted with a band round the fuselage in the national colours.

INDIA

1943-45

The Indian Air Force was formed on 1 April 1933 and normal RAF markings were used. In 1943, owing to confusion with Japanese markings, it was decided to outline the roundels with a thick yellow line, but this was not implemented, although a few aircraft may have been so marked. With effect from 24 June 1943 South East Asia Command markings were used, which initially dispensed with the red centre spot. Later a small roundel and fin marking of two shades of blue was used.

The force became the Royal Indian Air Force in March 1945 and reverted to standard RAF markings. On independence in 1947 Indian aircraft used the blue-on-white Buddhist wheel of life, the 'dharma chakra', as a wing and fuselage marking. The fin flash used the Indian national colours of saffron yellow, white and green. In 1948 roundels of the national colours replaced the wheel.

It is believed that an Indian National Air Force was being organised by the Japanese to fight the 'occupying colonial power', but if this existed there is no actual record.

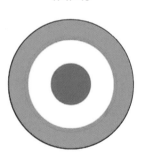

1947-48

The Indian 'chakra wheel' insignia on a Spitfire XIV in 1948. Aditya Krishna

Indian MiG 29. Jozef Grego

INDONESIA

During the Japanese occupation it is believed that anti-Dutch, pro-Japanese units adopted an aircraft marking of a red square with a black border, but this has to be confirmed. After the Japanese defeat in 1945, the former Dutch East Indies became a battleground between the colonial power attempting to regain control and the Indonesian nationalists. The Indonesian People's Security Force, Aviation Division, was formed on 9 April 1946, and the motley collection of mostly ex-Japanese aircraft bore the nationalist markings of red and white. This entailed painting over the bottom half of the Japanese roundel, or covering it with the nationalist flag of red over white. Former Dutch aircraft carrying the flag insignia simply had the blue portion painted out.

With the full independence of Indonesia on 27 December 1949 an air force was formed. Various red and white markings were used, culminating in a plain two-colour roundel and fin flash. By 1954 this had been changed to a white pentagon with a red border, the national flag being retained as a fin flash.

Navy and army aviation units were formed in the 1960s. Naval aircraft carry a black anchor on the white pentagon, army units a yellow or black five-pointed star. Current low-visibility markings feature a black outlined pentagon.

Right: *An Indonesian Army C-47.* via Kevin Curtis

Below right: *An Indonesian Navy DHC-5 Buffalo.* via Greg Kozak

Below: *Indonesian Lockheed C-130.* via Kevin Curtis

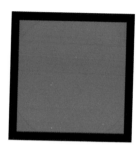
1942-46

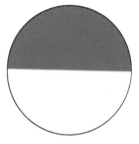
1946-49

1949-54

Army

Navy

IRAN

Tail, 1922-30

When the Imperial Iranian Air Force was founded in the 1920s a version of the national flag, green over white over red, was used as a wing and fuselage marking, and the rudder was painted in the same colours. A roundel form was soon adopted, with red in the centre. In 1970 the air force became the Islamic Republic of Iran Air Force, but the markings were not changed except for the addition of Islamic symbols to the national flag and therefore to the fin flash.

The Iran Imperial Air Force Aerobatic Team.

Islamic Republic of Iran Air Force Lockheed C-130. via Greg Kozak

IRAQ

1931-58

Tail, 1958-62

With a very few slight modifications, the wing and fuselage markings of aircraft of the Iraqi Air Force did not change between its formation in 1931 and 2003. It has consisted of a green triangle with a black border, superimposed on which is a representation of the Arabic word 'Jaish' ('army') in red. It is in the fin and rudder markings that the many political changes in the country have been mirrored. Between 1931 and 1958, when the monarchy was overthrown, the rudder was marked with vertical stripes of the four national colours; green, white, red and black; this later became a fin flash. The fin marking in use between 1958 and 1962 was vertically striped in black, white and green, with a red eight-pointed star on the white. In 1962 a horizontal striping of red, white and black was used. Later three green-outlined eight-pointed stars appeared on the white, then after 28 August 1963 these became solid five-pointed stars; initially these represented the proposed union of Egypt, Syria and Iraq, which never took place.

During the Gulf War of 1991 the inscription 'Allah u Akbar' ('God is great') appeared on the flag and the fin flash.

Since the formation of the new air force in 2004 aircraft have borne only the national flag; a roundel of the national flag was proposed, but so far never used. A new roundel, adopted in 2011, is red, white and red, the outer ring inscribed 'IRAQI AIR FORCE' in black. The emblem of the air force is superimposed over all.

1958-62

Tail, 1962-63

1962-63

Tail, 1963

1963

2013

An Iraqi MiG 21, 1985. via Greg Kozak

KURDISTAN

The semi-independent Kurdistan Provisional Government obtained some light aircraft in 2012, which bore the national flag at least below the wings.

IRELAND

Ireland's military aviation commenced with the establishment of the Free State in 1921, and former RAF aircraft were marked with the new national flag of orange, white and green. In 1922 a roundel form in national colours was used on the wings and fuselage, and as a vertical striping on the rudder. By 1923 a standard insignia had been adopted, which consisted simply of stripes across the wings and rudder in the national colours.

Irish ex-RAF Bristol F2B, 1922. Irish Air Corps

1922-23

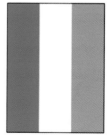

1923-39

1939-54

Ireland was neutral in the Second World War and used a Celtic boss insignia in green and orange on the fuselage and on top of the wings; stripes under the wings were retained. On a dark background the fuselage insignia was placed on a white square. In 1954 the Celtic boss was added to the white. The flag is also often used on helicopters.

Hawker Hurricane in 1945. Irish Air Corps

Aerospatiale Alouette III. Joe Maxwell

ISRAEL

The Jewish symbol of the Star of David has been used on Israeli military aircraft since the state's independence in 1948, applied to the wings and fuselage, almost always on a white disc. No rudder markings have been used except for a short time in the 1960s when red and white diagonal stripes were employed.

Israeli Douglas A-4 Skyhawk. Andreas Zeitler

ITALY

Italy was one of the first countries to use aircraft in war, in 1911. Up to 1914 the aircraft often used the arms of the House of Savoy as a fin marking, then between August 1914 and the entry of Italy into the First World War in April 1915 markings consisted of a simple black ring and various combinations of black stripes. From 1915 the underwings of aircraft were painted red on one side and green on the other. With the full involvement of Italy in the war, markings followed the Allied pattern of a roundel and rudder striping in red, white and green. Between 1918 and the Fascist government of 1922, roundels were sometimes used but the basic marking was rudder striping with the arms of Savoy on the white area.

The Fascist regime used its emblem – the fasces (a bundle of brown wood around a silver axe head on a blue disc) – on the fuselage. In 1940 Italy joined with Germany and declared war on the Allies. The rudder striping was painted out and a simple white cross bearing the arms of Savoy was marked across the fin and

1912-14

1914-15

Bleriots during the Libyan Campaign, 1912. Roberto Gentilli

1915-22

Fin, 1922-40

1922-40

Fin, 1940

Fin, 1940-43

1911 Italian Bleriot XI with the Savoy eagle on the fin. Roberto Gentilli

rudder, and the small fasces emblem was retained on the fuselage. Each wing was marked with a black circle surrounding three black fasces on a white or clear background. Occasionally the colours were reversed, with white fasces on a black disc. Some captured French aircraft featured the fasces across the original roundels.

Italy surrendered in 1943 and split into two separate areas, each with its own air force. In the south of the country, occupied by the Allies, the Co-Belligerent Air Force was formed. This used a red, white and green roundel and no fin markings. Some US-supplied aircraft retained the bars to the roundel. In the German-occupied north, the Italian Socialist Republic adopted a square version of the fasces insignia. The Italian flag was marked on the fuselage and fin.

After the war the Co-Belligerent markings continued in use for all Italian aircraft. From 1964 Italian Navy aircraft were marked with a black anchor in a black circle in addition to the roundel. The need for low-visibility markings has seen the introduction of a smaller roundel with a thin white area.

Aircraft of the Independent Sovereign Military Order of Malta, based in Rome, have occasionally carried the Order's cross insignia on the fuselage and even replaced the wing roundel as well.

Hanriot HD-1, 1919. Roberto Gentilli
1930 Ansaldo AC-3. Roberto Gentilli

1940-43

1940

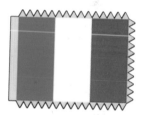

Fin, Italian Socialist
Republic, 1943-45

Italian Socialist Republic,
1943-45

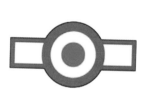

1943

1944-89

Independent Sovereign
Military Order, Malta, 1947

Independent Sovereign
Military Order, Malta, 1947

Navy

Second World War Caproni Ca-133 with the alternative style on the wings and rudder. Roberto Gentilli

Republica Sociale Italiano Fiat G.55 during the war in 1944.

IVORY COAST

This former French colony gained its independence in 1960 and formed its air arm in 1961. Aircraft are marked with a roundel and fin flash in the national colours of green, white and orange.

An Ivory Coast Beechcraft Bonanza F-33. via Kevin Curtis

JAMAICA

The air wing of the Jamaica Defence Force was formed in 1963. Markings are roundels in the national colours of green, yellow and black.

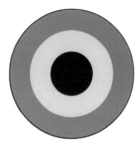

Jamaican Beech King Air.
via Greg Kozak

JAPAN

A start in military aviation in Japan was made before 1914, but the present Army and Navy Defence Forces date from 1954. Japanese military aircraft have always used the same marking, a red disc representing the sun, outlined in white or yellow if necessary on a dark background and, for Second World War Home Defence units, on a white square. In 1945, to facilitate surrender agreements, the Allies decreed that all Japanese aircraft should have their camouflage removed and carry a green cross in the normal insignia positions.

1915

Japanese Kawasaki T-4.
Jan Jorgenson

1944-45

1945

JORDAN

Formed as the Arab Legion Air Force in 1949, and becoming the Royal Jordanian Air Force in 1956, this air arm's insignia has not changed. It consists of a black, white and green roundel with a red sector at the top; the latter contains a white seven-pointed star representing the first seven verses of the Koran. The national flag is used as a fin flash.

Jordanian Lockheed F-104 Starfighter. Andreas Zeitner

KAZAKHSTAN

This large central Asian republic became independent in 1991, and a new flag of a blue disc bearing a yellow sun and eagle was soon adopted as the insignia for government aircraft, only appearing on the fin. Kazakh military aircraft bear a yellow-bordered red star, superimposed on which are a yellow sun and an outline yellow eagle.

1991

A Sukhoi Su-27 in Kazakhstan. Andreas Zeitler

KENYA

DHC-2. via Kevin Curtis

An air unit of the Kenya Defence Force was formed in 1933, but did not become a viable military unit until 1939. Its aircraft carried standard RAF markings, and the unit was disbanded in 1945. After independence in June 1964, a Kenya Air Force was established. Aircraft are marked with the Kenya national colours of black, red and green, and in roundel form each colour is separated by a thin white line.

KOREA DPR (NORTH KOREA)

The Korean People's Armed Forces Air Force had its origins in 1945, and became fully fledged in 1948. The insignia is a red star surrounded by thin rings of the Korean national colours, blue, white and red. This roundel is usually applied on the fuselage and under the wing; on some early aircraft it was also carried across the fin and rudder.

1920

North Korean MiG 15.
George Trussell

KOREA (ROK)
(SOUTH KOREA)

1950

An attempt was made to form a Korean Air Force, based in the United States, in the 1920s, at a time when the country was part of Japan. Some aircraft were purchased and marked with the red and blue 'yin-yang' symbol and the letters KAC (Korean Air Corps), but neither aircraft nor crews ever reached Korea. Korea became independent in 1945 but was immediately split between the communist north and the US-occupied south. The Republic of Korea Air Force was just being organised when the North Korean forces invaded. Some former Japanese aircraft were obtained by South Korea and identified as Korean by painting the 'yin-yang' symbol over the red disc of Japan. Most South Korean aircraft were supplied by the United States and the insignia reflected this, although a red and blue 'yin-yang' replaced the US white star. During the Korean War most ROKAF aircraft carried a large black 'K' on the fin. Recently three black lines on each side of the roundel replaced the US-type bars. Greyed-out low-visibility markings are used by many combat aircraft.

South Korean KA-1.
Andreas Zeitler

KUWAIT

A roundel in the pan-Arab colours of green, white, red and black, and a national flag fin flash, were used by Kuwaiti aircraft between the formation of the air force in 1961 and the Gulf War of 1991. Very occasionally the word 'Kuwait' in Arabic was used, in white, on the outer green part of the roundel. During the Gulf War aircraft bore the legend 'FREE KUWAIT' and the fin flash, and since 1991 this has been the only marking.

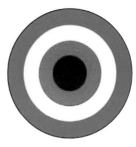

1961-91

Kuwaiti De Havilland Dove with roundels. via Kevin Curtis

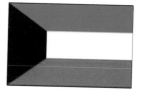

1991

A Lockheed C-130L100 in 2001. Anthony Osborne

KYRGYZSTAN

This central Asian republic became independent in 1991, and has a small air force. This insignia is a yellow design representing a yurt, a traditional nomadic tent, on a red disc.

Kyrgyzstan Mil-24.

1955-75

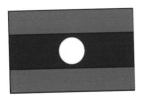

LAOS

Laos became independent in 1952 and formed its first air arm in 1955, which became the Royal Lao Air Force in 1960. After a long period of civil war the monarchy relinquished control to the Pathet Lao and the country became the Lao People's Democratic Republic. Between 1955 and the fall of the monarchy in 1975 the air force roundel was a red disc bearing three white elephants under a parasol standing on five steps, representing the five First Commandments of Buddhism. Since 1975 the national flag has been a horizontal tricolour of red, blue and red, with a white spot on the blue, and is used as a wing marking; the fin version is often slanted.

A Royal Lao Air Force C-47, 1970. via Kevin Curtis

A Laotian DHC-2.

1918-40

1918

A Latvian An-2 carrying current markings.
Jan Jorgenson

LATVIA

This Baltic republic was independent from 1918 to 1940, and again from 1990. Initially Latvian aircraft were marked on wing tips and rudder with the national colours of red, white and red. This was soon changed to a red swastika turned on its point, usually on a white disc on fuselage sides and the bottom of the wings. On some aircraft the markings were omitted on the top wing, or shown without the white disc, but this was later amended.

During the late 1930s the paramilitary National Guard marked its aircraft with a red indented cross on wings and fuselage. Only naval aircraft of this period carried any fin or rudder marking, which was a black rope-and-anchor design. A Latvian unit attached to the German Air Force saw service on the Eastern Front, and its aircraft carried a black cross on the wings and fuselage, and a reversed swastika on the fin.

The new Latvian Air Force carries the national flag, maroon split into three parts with a thin white stripe as a fin flash; the wings and fuselage carry a roundel version. From 1994 the Latvian Army Reserve used a red outline saltire-type cross.

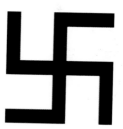

Tail, 1941-44

1941-44

Army Reserve

National Guard, 1937-40

Navy, 1936-40

LEBANON

A Lebanese Cessna 208. via Greg Kozak

A DH Vampire T55.

Lebanon secured its independence, after being a French mandate, in 1943, and formed an air force in 1949. Its aircraft have always carried the same markings: the national flag as a fin flash, red over white over red, with a green cedar tree on the white, while the wing and fuselage roundels are in the same colours but consist of a green spot on a white triangle on a red disc. On some early aircraft this roundel is repeated on the fin.

LESOTHO

1980-86

Formerly known as Basutoland, this land-locked country is surrounded by South Africa and achieved its independence in 1966. The Air Squadron of the Royal Lesotho Defence Force was formed in 1978 and carried the national flag, which featured a 'mokorotlo', the national hat, and a roundel in the four colours of the flag. The flag changed in 1987 and again in 2006. The roundel colours have used green, white and blue since 1987.

LIBERIA

The Air Reconnaissance Unit of the Liberian Army was founded in 1976 and bore the national flag on the fin. From 1986 the unit bore an insignia on wings and fuselage very similar to the USA style but with a much smaller white star. Currently only the national flag is used.

LIBYA

Right: *A 1976 Dassault Mirage 5.*

Far right: *A Free Libya MiG 21U, 2011.*
via Kevin Curtis

1959-69

Above: *A Libyan 1968 Lockheed T-33.*
via Kevin Curtis

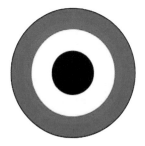

1969-78

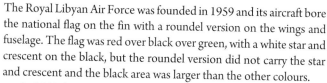

A MiG 23 with current markings. via Greg Kozak

1978

The Royal Libyan Air Force was founded in 1959 and its aircraft bore the national flag on the fin with a roundel version on the wings and fuselage. The flag was red over black over green, with a white star and crescent on the black, but the roundel version did not carry the star and crescent and the black area was larger than the other colours.

In 1969 the monarchy was overthrown and the Libyan Republic Air Force changed to a flag, roundel and fin marking in red, white and black, exactly the same as Egypt, as the two countries were forging very close ties at this time. 1978 saw political conflict, with Egypt and Libya leaving the Federation of Arab Republics, and a new flag was adopted in plain green to denote complete devotion to Islam; this was reflected in the plain green roundel and fin marking.

Following the revolution of 2011 the flag was changed to that of 1951, and roundels were split horizontally red over black over green with a white star and crescent on the black. An alternative roundel is an outer ring of red on top and green below separated by a thin white line; the centre is black with a white star and crescent.

2012

A Mil-35, 1990.
Chris Lofting

2013

LITHUANIA

1919-20

This southernmost of the Baltic republics gained its independence in 1919. The national colours of yellow, green and red were marked on the fuselage, rudder and wings in a diamond pattern, and the central red area was later marked with a representation of a mounted figure of St George, in white. From 1920 aircraft began to be marked with an ancient Lithuanian symbol, the white double cross of Vytis. At first this appeared on a red shield, but by 1921 it was normally used plain, with a black border on pale surfaces.

The Lithuanian National Guard was formed in 1938, and its aircraft were marked with the double cross insignia under the wings and on the rudder, in white on a red shield.

Lithuania became part of the Soviet Union in 1940, but regained its independence in 1990. A return was made to the Vytis cross, outlined in black where necessary. The National Guard carry its own insignia on the fuselage and occasionally on the wings, with a cross marking on the fin.

1920-21

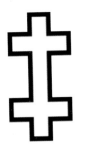

1921-40

Central Lithuania, wing, 1920-21

A Lithuanian An-26. Jan Jorgenson

Central Lithuania, wing, 1920-21

National Guard, 1936-40

National Guard

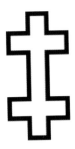

CENTRAL LITHUANIA

In October and November 1920 Polish forces set up a puppet state called Central, or Srodkowan, Lithuania. Aircraft, actually of the Polish Air Force, carried special markings that were unusual in that they differed on the port and starboard wings. On the port wing a red square with a white border was carried, while on the starboard the colours were reversed. No fuselage markings were used and the rudder was split vertically red and white, with red leading, the reverse of the then Polish practice.

LUXEMBOURG

In the late 1950s the Luxembourg Defence Force acquired a small number of light aircraft. Although civil-registered they carried roundels of blue and white with a red lion on the central white. A fin flash of either the national flag or a design of red, white and blue diagonal stripes was used. The Luxembourg Defence Force no longer possesses an aviation complement, but for political reasons NATO aircraft are registered in the Grand Duchy. The NATO symbol, a four-pointed compass design in blue, appears in normal positions on wings, fuselage and fin. Since the mid-1980s aircraft have also carried a roundel based on the national arms of Luxembourg, a red lion on a blue and white striped disc.

Tail, 1957-68

1957-68

Right: *A Luxembourg Piper Cub, 1955.*

Below: *A Luxemburg NATO Boeing E-3 Sentry.* Anthony Osborne

NATO

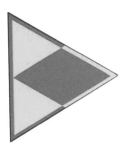

MACEDONIA

This republic, once a part of Yugoslavia, obtained its first aircraft in 1994, and the insignia, a triangular design of red and yellow based on the national flag was first used in 1998. The national flag is used as a fin flash.

A Macedonian Zlin 242.
Zoltan Rados

MADAGASCAR

This former French colony, a large island in the Indian Ocean, formed its military aviation unit in 1961. The national flag of red and white for the Hova tribe, and green for the Indian Ocean, was used as a fin flash, and in roundel form on wings and fuselage. Often only the fin flash is used.

MALAWI

Malawi, once called Nyasaland, founded its army air wing in 1966, two years after independence. Two flags are used as markings: the national flag, of black, red and green, and the army flag of red, white and red, with a white diamond. A roundel version of the national flag is in use.

1966

A Malawi Douglas C-47. Winston Brent

MALAYSIA

Military aviation in this region dates from the formation of the Straits Settlements Volunteer Air Force in March 1936. This became the Malayan Volunteer Air Force in September 1940, and took part in the campaign of 1941-42.

On 1 October 1950 the Malayan Auxiliary Air Force was established, and this became the Royal Malayan Air Force on 1 June 1958. All these forces used standard RAF markings.

With independence and a change of name to Malaysia, the Royal Malaysian Air Force was founded, with new markings, on 1 June 1963. The markings consisted of a pale blue square with a dark blue border, in the centre of which is a yellow fourteen-point star, representing the country's fourteen states; there is also a fin flash in these colours. In 1982 the square became a roundel, and low-visibility requirements have resulted in much smaller markings.

1963-82

Fin

Navy

Above: *The square insignia on a Malaysian CAC 27 Sabre.* via Kevin Curtis

Right: *Malaysia's current insignia on a Northrop F-5.*

Below: *Malaysian Navy insignia.*

MALDIVES

The small air component of the Maldives Defence Force uses a roundel of red and green with a white crescent in the centre.

A Maldives HAL Dhruv. via Greg Kozak

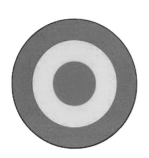

MALI

The pan-African colours of red, yellow and green have been used on Mali's aircraft since the formation of its air force in 1961. Recently a smaller roundel has replaced the fin flash or rudder striping of the same colours.

A Mali Douglas C-47.
via Greg Kozak

MALTA

1973-80

Although helicopters were acquired for the Malta Land Force in 1972, it was not until April 1973 that they came under the authority of the First Regiment of the Armed Forces of Malta. They then carried the insignia of the regiment, a white number '1' superimposed on a roundel of red and blue.

A Maltese CASA 212 Aviocar.

In April 1980 the aircraft became part of the Malta Task Force and adopted a white-over-red roundel bearing the black letters 'TF'. In 1988 military aviation reverted to the command of the First Regiment and the 1973-80 insignia.

1980-88

In 1992 a red and white roundel with a black representation of the George Cross in the centre was adopted, with the national flag as a fin flash.

MAURITANIA

Since its formation in 1961, the aircraft of the Mauritanian Air Force have been marked with the national flag, green with a yellow star and crescent. This was originally in a square format, but a roundel version has appeared since the late 1980s.

MAURITIUS

Various government aircraft of this small Indian Ocean island have recently adopted a roundel version of the national flag, which is red, blue, yellow and green.

MEXICO

Mexican aircraft have always worn a representation of the national colours of red, white and green. It is not generally known that Mexico was one of the earliest countries in the world to use aircraft for military purposes, dating back to 1911. The air corps was formed in 1914 and by 1915 had adopted a shield marking in the national colours. In about 1920 this was changed to wing striping or standard roundels. A triangular insignia for wings and fuselage became usual by 1922, together with rudder striping. The Mexican Expeditionary Force in the Pacific in 1945 used these markings alongside American ones.

In recent years the Mexican Navy has obtained aircraft and marked them with either one or two crossed anchors in either black or white behind the triangles.

After the Second World War there was a suggested change to a roundel split vertically with the national colours, but it is very unlikely that this was ever used.

1915-20

1920-22

Mexican Navy Beech Baron 58s.
via Kevin Curtis

Navy

1920

MOLDOVA

This former Soviet republic, bordering on Romania, formed an air force after independence in 1991. Aircraft bear a red eight-pointed star edged with yellow and blue on a white disc, colours that were chosen as Moldova has close ties with Romania.

A Moldovan Antonov An-2. Chris Lofting

MONGOLIA

1936-45

Tail, 1936-92

Mongolian MiG 17s.
via Greg Kozak

Mongolia obtained its first military aircraft in 1925, supplied by the Soviet Union, and they probably carried normal Soviet red stars. By the mid-1930s these aircraft bore a traditional Buddhist ideogram called the 'zoyombo', which comprised many elements including bravery, friendship, and the symbolism of the yin and the yang. This zoyombo was marked on the fin and sometimes on the fuselage, and was either red or yellow depending on the background colour. Soviet red stars were often carried in addition. Aircraft marked in this way took part in the conflict with Japanese forces between 1938 and 1945.

After the war and the disruptive Chinese civil war, Mongolia became even more within the Soviet sphere and aircraft tended to carry a simple red star, possibly surrounded by a thin red ring. The zoyombo returned in the early 1960s, usually marked in red on the fin and yellow on a red star on the wings, and included a Soviet star. Since 1992 the star element of the zoyombo has been discarded.

Wing, 1936-92

　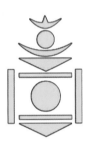

1992

MONTENEGRO

2103

Previously a republic within Yugoslavia, Montenegro became fully independent in 2007. Initially its military aircraft carried a red disc with a yellow border and the national flag, but the insignia was changed in 2011 and now consists of a version of the yellow cross of St George, an ancient Balkan symbol. This is surrounded by a red ring and features side bars of three red lines.

Below left: A Montenegran 2007 Aerospatiale Gazelle.
Joop de Groot

Below: A 2013 Soko Super Galeb.
via Greg Kozak

A Moroccan Max Holste Broussard. via Greg Kozak

MOROCCO

Since its formation in 1956, the Moroccan Air Force has used the national flag of red with an open green five-pointed star known as Solomon's Seal as a fin flash. The wings and fuselage marking is a roundel version, the star being topped with a green and yellow crown for the monarchy.

The Naval Gendarmerie used a roundel of red and blue with a black anchor, topped with a red star. The Moroccan Navy uses a normal roundel backed with two crossed anchors.

Gendarmerie

MOZAMBIQUE

This former Portuguese colony became independent in 1975 and had formed an armed combat air force by the following year. A roundel based on the national flag has been used on the wings and the fin, and consists of a red triangle on a black disc; the yellow markings represent a book for education, a hoe for agriculture, and a rifle for the fight for independence. Occasional use of this has been noted as a fin flash.

MYANMAR

Previously known as Burma, this nation became a separate country within the British Empire in 1937 and established a small volunteer air unit before the Second World War. Aircraft bore normal RAF markings with distinctive serial numbers. This unit was disbanded in 1942. The country became fully independent in 1948, and formed the Union of Burma Air Force in 1955. Unusually the national colours are not used as a basis for the air force insignia; the colours are blue, white and yellow, as featured on the air force ensign, and are displayed on the wings and fuselage in triangular form and on the fin as a square flash.

A Myanmar Lockheed T-33. Tim Spearman

NAMIBIA

This country, once known as South West Africa, became independent in 1990. A defence force was formed, and initially the aircraft did not carry insignia, just the national flag on the fin and the letters 'NDF' and a serial number. Since 2011 Namibian Air Force aircraft have carried the national arms with wings and supporting wreaths.

A Namibian Cessna 337. via Greg Kozak

NEPAL

Tail

The air wing of the Royal Nepal Army was formed in the mid-1960s, and became the Royal Nepal Air Force in 1979. Aircraft are marked with a red six-sided star bearing a black trident. The fin flash is the uniquely shaped Nepalese national flag comprising two joined right-angle triangles.

Left: *An HS-748.* Aditya Krishna

Below: *An Aerospatiale Super Puma.* Matias

NETHERLANDS

A Fokker D.21; the orange triangle is a sign of the Netherlands' neutrality in 1939.

A Gloster Meteor F-8, 1948. via Kevin Curtis

1914-21

Military air activity in the Netherlands began with an army balloon unit in 1886. An aviation unit of the Royal Netherlands Army was founded on 1 July 1913, and by December of that year an orange disc was painted under the wings. Immediately after the start of the First World War, on 5 August 1914, it became imperative to demonstrate the country's neutrality, so orange discs were painted on top of the wings and the rudder. By a decree of 17 April 1917 the discs were also marked on the fuselage and the rudder was orange. On 11 June 1921, some eighteen months after its introduction in the East Indies, a new insignia was decided, which used a segmented roundel coloured red, white and blue, with an orange spot in the centre; the rudder was painted horizontally in the main colours.

1921-39

With the start of the Second World War the Netherlands once again wished to affirm its neutrality. Once again orange was adopted, officially from 23 September 1939, and took the form of an inverted triangle with a black border. Germany occupied the country in May 1940. Exiled Dutch forces operating from Britain used aircraft with the RAF marking, but with an orange and black triangle next to it, on the fin.

In 1946 the segmented roundel was reintroduced with a smaller orange spot, and the national flag as a fin flash; the latter was discontinued from about 1959. Low-visibility requirements produced a smaller insignia, and later a greyed-out version.

1939-41

A Netherlands Aeronca L-4 carrying the national flag, which was often used on the fuselage and wings, c1945-50. The badge near the front is that of the Artillery Training School.

The low-visibility version of the Lockheed C-130.
Anthony Osborne

NETHERLANDS EAST INDIES

Some Dutch aircraft were sent out to the East Indies in 1914 but did not seem to use any markings. It is possible that they were

1942-45

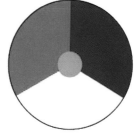

1945

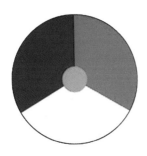

left on the aircraft after operation in the home country but may have been erased to avoid confusion with Japanese roundels. The Indies started using the segmented insignia from January 1920, but with the Japanese threat of 1941 it was replaced with the orange triangle.

Following the Japanese occupation, Dutch forces that had escaped to India or Australia used a rectangular version of the national flag, which was officially recognised from March 1942. After the war aircraft used either Allied markings with the addition of an orange triangle or the rectangular flag, or even the segmented insignia. Dutch forces withdrew from the area on the independence of the colony as Indonesia in 1949, but kept a presence in New Guinea until 1963.

NETHERLANDS ANTILLES

The Netherlands West Indies Defence Force was active in the Caribbean area between 1940 and 1945, and its aircraft used the segmented and later the flag variations.

NEW ZEALAND

Several attempts had been made to organise a military aviation unit before the First World War, but the first official New Zealand Air Force was formed in 1923, and was given the prefix Royal in February 1934.

Until 1942 New Zealand followed RAF practice in aircraft markings, but during the war with Japan many variations were used. The central red area was made progressively smaller, until it was merely a dot. The central spot colour was then changed to blue and, although not as small as the red spot had become, it was still much smaller than in the standard RAF roundel. Many New Zealand aircraft used in this period were supplied by the United States, and often carried bars attached to the roundel; these were outlined in blue on pale surfaces. Fin flashes from 1942 to 1945 became very thin.

After the war a return was made to British markings. In 1957 a specific New Zealand marking was devised, with a white fern placed on the central red of the roundel. Because it was said that it looked like a white feather (a symbol of cowardice) it was soon changed to silver. Unfortunately, on natural metal aircraft it now

Tail, 1942-44

1942-44

New Zealand's fern-type insignia on T-6 Harvards. via Kevin Curtis

The standard current New Zealand markings on a Fokker F-27 Friendship. Paul Adams

A low-visibility roundel on a Douglas A-4 Skyhawk. Paul Adams

1944-45

1957-70

1970-89

1989

looked like worn paint, so from 10 October 1970 a red kiwi replaced the central spot. Recent low-visibility markings consist of a red kiwi on a blue disc or even a greyed-out version. Throughout this period RAF fin flashes have been used.

NICARAGUA

Above: *A Nicaraguan Douglas C-47, 1980.*

Right: *This Mil-17 carries the Sandinista roundel.*
via Greg Kozak

1936-42

1942-62

1942

The turbulent recent history of this Central American republic has been mirrored in the frequent changes of aircraft markings. After several false starts in the 1920s, an aviation corps seem to have been organised by 1927, and aircraft carried the national coat of arms on the fuselage. It was 1936 before more military aircraft reached Nicaragua. Now known as the National Guard Aviation Corps, they bore red, white and blue horizontal rudder striping.

From the mid-1940s to the mid-1950s the fuselage roundel was coloured blue, yellow and red, and the wing markings were a white triangle on a blue disc with a red border. The rudder was marked with seven horizontal blue stripes on white with a thin yellow stripe next to the hinge.

From about 1962 the fuselage roundel was changed to the insignia of the air corps, a blue disc with a red border and complex markings in yellow and red. The stripe next to the rudder hinge was changed to red, but this was very gradual, with both colours used for many years, and some aircraft used just the blue and white stripes. Up until the 1980s the triangle-type markings continued to be used on the wings.

Tail, 1962-82

1962-82

Tail, 1982-90

With the victory of the Sandinistas in 1979 the markings changed yet again. Aircraft bore the legend 'Fuerza Aerea Sandinista', and on the fin the letters 'FAS' and a red and black flash. By the late 1980s a new roundel in orange with a red and black design and a central orange star was in use. Some aircraft also carried the blue/white/blue national flag as a fin flash.

Since the electoral defeat of the Sandinistas in 1990, this last roundel has gradually disappeared. It would appear that future markings will simply consist of the national flag as a fin marking or placed on a helicopter.

1990

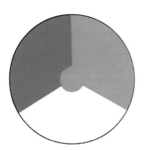

1961-80

NIGER

This former French colony established its air force in 1961. The national flag, which has always been used as a fin flash or rudder marking, is orange for the deserts of the north, white for purity, and green for the forests of the south; there is an orange disc in the central white area. Initially the roundel was segmented but it is now divided horizontally.

1961-80

A Niger Dornier Do-228.

NIGERIA

1964

The Nigerian Air Force was formed in 1964 and has always used the national colours of green and white as a roundel and fin flash. The fin has often carried a roundel instead of, or as well as, the national flag.

Navy aircraft now carry a black anchor superimposed on the roundel and fin flash. A modified version of the roundel was introduced in 2009, although the previous marking is still in use.

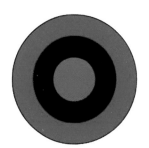

Biafra, 1967-70

Above: *The normal roundel on a Nigerian L-39AZ.* Ken Iwelumo

Above right: *The current Nigerian roundel on a Mil-24.* Ken Iwelumo

Right: *A Nigerian Navy Agusta 104.*

Navy

BIAFRA

Between 30 May 1967 and January 1970 the Republic of Biafra was independent of Nigeria and for the whole period fought a bitter war with its larger neighbour. Biafra's few aircraft were marked with the national colours of red, black and green in roundel form, and in flag form on the fin. Some aircraft used the flag as a fuselage marking.

NORWAY

Norway first acquired military aircraft in 1912, and from about 1915 up to 1945 the army and navy had separate forces. They did, however, share the same markings, which were stripes across the wings and vertical on the rudder in red, white and blue in the proportions of the national flag.

After the fall of Norway in 1940, many Norwegians escaped and were trained in Canada and the United States. The British Norwegian squadrons carried a small national flag in addition to standard RAF markings and, from about 1944, pre-1940 rudder striping. Aircraft in North America usually carried full Norwegian markings, while those on operational use bore a combination of US and Norwegian insignia. For a brief period after the liberation a roundel similar in colours and order to that of France with the thin white portion was used on 'liberated' German aircraft.

A Royal Norwegian Air Force was established on 21 November 1945, and a roundel was designed consisting of a blue ring around a triangular representation of the national flag. No tail markings were carried. Recent changes to low-visibility markings have produced a black outline form of the insignia.

1915-45

A Norwegian Fairchild M-62 showing pre-1945 markings. Jan Jorgenson

This Westland Lynx carries the current roundel. Jan Jorgenson

OMAN

The Sultan of Oman's Air Force was first established in 1959, at which time the Omani flag was plain red; the first markings were a red disc, usually outlined in white. The arms of the country comprised crossed swords, and was applied in red on a white shield as a fin flash; this was changed in the 1960s to a white-on-red fin marking. In 1970 a new flag was adopted, which included the arms and a green area. The fuselage and wing roundels were red with a green outline and carried the crossed swords in white. By the end of the 1970s many aircraft dispensed with roundels and just bore a white-on-red shield.

From 1983 new colours were adopted for markings, and unusually they were not featured on the national flag; they comprised white or yellow swords on a blue shield, and some aircraft bore these on the wings but most carried them only on the fin. Larger passenger-type aircraft often carried the full national flag on the fin.

An Omani Britten-Norman Islander. BN Historians

Tail, 1959-65

Wing, 1959-65

Tail, 1965-70

Wing, 1965-70

1970-83

1977-83

PAKISTAN

Pakistan was formed with the partition of India in 1947, and the air force consisted of a portion of the original Indian Air Force. The initially proposed national flag was plain green with a white star and crescent, and the air force insignia was to be a green disc, also with the star and crescent. However, mindful of non-Muslim minorities,

1947

Navy

a white band was added to the flag and a green and white roundel became the air force marking, normally with a thin yellow outline. However, there is no evidence that this flag and roundel were ever used. The fin flash has always been a green square with the white star and crescent. Aircraft of the Pakistan Navy have featured a black anchor across the roundel.

A Pakistani Cessna T-37. via Kevin Curtis

Tail, 1931-39

PANAMA

Panama's first attempt at forming an air arm was in 1931, when aircraft bore a red and white roundel with 'R de P' in blue in the centre. A blue and white roundel with the inscription 'GN 31' was carried on the fin; presumably this marking stood for 'Guarda Nacional 1931', and it continued in use through the 1930s.

Wing, 1931-39

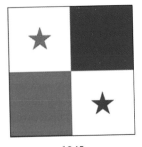

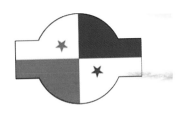

1945

Military aviation virtually disappeared by 1939, possibly because the large United States element guarding the Panama Canal made an independent force unnecessary. Another attempt was made after the Second World War, and some aircraft bore just the national flag on the fin.

The first modern Panamanian military air arm was formed in 1964. Aircraft were marked with a roundel and side bars as a representation of the red, white and blue of the national flag. Rudders were painted red, white and blue, with a red and blue star on each of the white areas.

PAPUA-NEW GUINEA

The Papua-New Guinea Defence Force was formed in 1974 and its aircraft carry a roundel form of the national flag, which is black, red and green with a yellow bird of paradise in the centre; this bird in flight is sometimes marked on the fin.

A Papua-New Guinea IAI Arava. via Greg Kozak

PARAGUAY

It was not until the late 1920s that Paraguay formed its first military aviation units, although aircraft had been use on both sides in the 1922 revolution – these may have been the first air combats in South America. Aircraft in the 1930s carried a red, white and blue roundel, or on the Fiat CR 20 diagonal, chord-wise wings. There were no fuselage markings and rudders were painted horizontally in red, white and blue with a yellow star on the white. The roundel and rudder markings have remained the same with few slight differences. Modern aircraft carry a fuselage roundel and a fin flash.

1929-38

A Paraguayan North American T-6. via Kevin Curtis

During the 1947 civil war government aircraft used stripes on the wings and fuselage to distinguish themselves from the rebels, who used roundel-marked aircraft. A crude black 'V' was painted next to the wing and fuselage roundels.

Paraguayan Navy aircraft of the 1950s bore black anchors instead of roundels and omitted the yellow star on the rudder. Modern aircraft have a black anchor marked over roundels.

Navy

PERU

Peru made its first attempts at military aviation in 1912, but it was 1930 before this became viable. The nation's colours of red and white have always been used in roundel form and as vertical rudder stripes. Around 1930 aircraft of the Peruvian Army had three white and three red stripes on the rudder, while more modern aircraft have used fin flashes. Navy aircraft have a black anchor incorporated into the roundel, while the army replaced the centre spot of the roundel with a triangle; more recently the army has used a standard roundel carrying a yellow sword.

Army, 2013

Army

Navy

Left: *A Peruvian Navy Grumman Tracker.* via Kevin Curtis

Below: *A Peruvian Army Antonov An-32.* Amaru

1935-41

PHILIPPINES

The Philippines Army Air Corps was established on 2 May 1935 while the country was still a dependency of the United States. Its aircraft were marked on the wings, fuselage and fin with a blue diamond bordered in blue and white. The corps was amalgamated with the US Army Air Corps on 15 August 1941 and was disbanded after the Japanese occupation.

Reformed as the Philippines Air Force on 3 July 1947, the diamond marking, now with an outer red border, was flanked by intricate white bars with blue borders.

A Philippines Britten-Norman Islander.
BN Historians

The same make of aircraft belonging to the Philippine Navy. BN Historians

POLAND

The national colours of red and white were adopted by Poland in 1831 and formed the basis of the national flag on independence in 1918. Officially the Polish Air Force became operational in November 1918, but owing to the chaotic situation prevailing in the area the three airfields of Warsaw, Lwow and Krakow adopted their own individual versions of a red and white insignia. Many ex-German aircraft had their black crosses obliterated by various versions of a red and white circle. Until December 1918 Warsaw had a red and white shield split diagonally, Lwow had red and white striped wing tips, and Krakow a red 'Z' on a white square. The official insignia was introduced on 1 December as a chequerboard design in red and white; this was initially in four plain squares, but soon acquired a contrasting red and white border.

During the period up to 1920 many British and French aircraft were obtained, and a large number retained their original markings, although many British aircraft changed the rudder striping to plain red over white, while some French aircraft bore French roundels on the port wing and Polish on the starboard. The insignia in general use by 1921, and the war with Russia and Ukraine, remained unchanged until the destruction of the air force in 1939.

Body, 1918

Tail, 1918

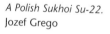

A Polish Sukhoi Su-22. Jozef Grego

Warsaw, 1918

Wing, 1918

Polish airmen who escaped to France were allowed to paint the Polish insignia over the French roundel on the fuselage side of their aircraft. Those that went on to form Polish squadrons in the RAF carried small versions of the square insignia in addition to British roundels

With the return to sovereignty in 1945, the pre-war insignia was reintroduced. Although Poland came within the Soviet sphere of influence, it seems to have resisted any moves to adopt the marking based on the red star. With the advent of democracy the Polish national marking remained unchanged. Aircraft of the Polish Border Guard use a circular version of the national insignia.

A Polish Border Guard PZL Wilga. via Greg Kozak

SILESIA

Conflict over Silesia in 1921 was resolved by a referendum that returned the area to Germany. The Polish-Silesian armed forces in this conflict were actually Polish regular troops. Aircraft of the 'Silesian Air Force' carried markings of a pale blue square with a black border on the port wing, and a white square with a black border on the starboard. Rudders were painted white, pale blue and black.

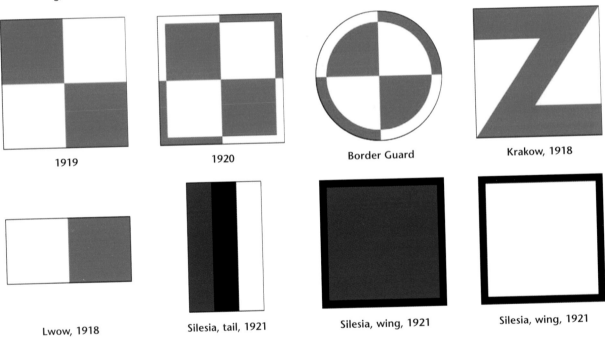

1919   **1920**   **Border Guard**   **Krakow, 1918**

Lwow, 1918   **Silesia, tail, 1921**   **Silesia, wing, 1921**   **Silesia, wing, 1921**

PORTUGAL

A Portuguese North American T-6. Carlos Oliveira

Portuguese military aviation dates from 1912, but the official army and navy air corps were established on 1 October 1916. Initially a red and green roundel and rudder striping were used, but by the time of the formation of the First Air Group on 17 December 1917 a special form of red and white cross was displayed on the wings, with the national flag as a rudder marking. The cross has since been marked on a white square, or far more frequently on a white disc. During the Second World War some aircraft dispensed with the white background.

The army and navy air arms amalgamated on 1 July 1952. Since then it has been usual to carry a simple red and green fin flash in the proportions of the national flag. Low-visibility markings have reduced the size of the insignia, which is now often used in a grey outline.

Tail, 1915-16

1915-16

A Lockheed P-3 Orion with low-visibility markings.
Carlos Oliveira

1916-52

QATAR

The air wing of this small Arab Gulf state was founded in 1968, followed by the Qatari Emiri Air Force in 1974. It has always used a roundel of maroon, sand and white, while the fin marking is a pair of national flags crossed on a black square.

*A Qatar Dassault Mirage
F1.* via Greg Kozak

ROMANIA

1913-15

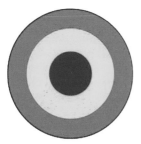

1915-41

Romanian military aviation dates from 1910, and many of the early Romanian flyers were very important in the history and development of aviation. Aircraft carrying roundels of red, yellow and blue saw action in the second Balkan War of 1913; the order of the colours was variable, perhaps as a result of the difficulty of determining colours from black and white photographs, but as many of the aircraft were French the only requirement would have been to paint the white part yellow. This insignia, together with a vertically striped rudder, continued in use until 1941.

Romania joined Germany in the invasion of Russia, so had to adopt an Axis-type cross marking; this was an indented yellow cross with a small roundel in the centre. There were many variations of this and many aircraft continued to use the rudder striping.

On 24 August 1944, with the Russian invasion of Romania, the country sued for peace and declared war on Germany. Markings immediately reverted to the roundel form, which was retained until after the formation of the communist republic in 1947. Romania

A Romanian Second World War IAR 80, 1943. S. Fopma

1941-44

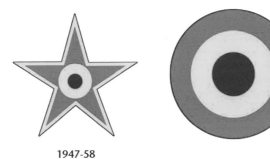

1947-58

then joined Hungary and Bulgaria by adopting a red star with the old roundel in the centre. In 1985, amid much political turmoil, the roundel form once again appeared, marked on wings and fuselage with a smaller version on the fin.

A Britten-Norman Islander in current Romanian markings. BN Historians

RUSSIA

The Imperial Russian Flying Corps was founded in 1912, as was the aviation element of the Imperial Russian Navy. Flying Corps aircraft were marked with a representation of the national flag, a horizontal tricolour of white over blue over red. Navy aircraft carried a blue cross of St Andrew on a white rudder in addition to the flag markings. By 1915 a roundel form was adopted, in red, blue and white; the central white area was often much larger than the other colours, which gave the impression of a white disc with a thin red and blue border. These roundels were marked on wings and fuselage, and often on fins, rudders and elevators.

During the 1917 revolution came a period of chaotic disintegration. Many units defected to the revolutionaries and their aircraft markings were based on the colour of the revolution – red. Initially roundels were painted over in red, but there were many variations, often featuring a red star; some were even noted with a red skull and crossbones. With the formation of the Workers and Peasants Air Fleet on 24 March 1918 the red star was the preferred marking, and this, with very little variation, has been used ever since.

Early Soviet naval aircraft carried a red or black anchor on the fuselage. From the early 1920s the red star was outlined in white or yellow if against a dark colour. There was even the earliest form of low-visibility markings – a black outlined version of the star.

When the Soviet Union returned to its original name, Russia, in 1992, red stars continued in use. Since 2009 the red star has been bordered with blue and white to match the original Russian colours. Low-visibility markings are a plain outline red star.

Russian naval aircraft carry a 'wavy' version of the blue-on-white St Andrews flag.

During the Russian Civil War of 1918-20 that followed the revolution, pro-Tsarist forces were still active. Many carried the Imperial roundel and some the cross of the Russian Orthodox Church. British, French and American forces, among others,

1915-16

Navy, 1915-17

1915-17

*A Russian Imperial Air Corps 1916 Nieuport 17.
Yefim Gordon*

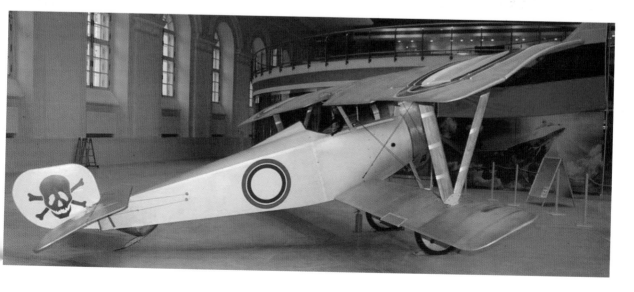

1917-22

1918-

Don Republic, 1918-21

1918

Chechnya, 1993-94

A 1920 Workers and Peasants Air Fleet Nieuport 17.
Yefim Gordon

intervened on the side of the White Russians and used their own insignia, but some used pre-revolutionary roundels and fin striping to match. The many anti-communist groups included the following:

The All-Greater Forces of the Don Air Corps A white disc on fuselage and wings with a triangle in either black or dark blue. The vertical rudder striping was, from the front, white, blue and red.

The First Siberian Air Detachment A rectangle of white over green.

The Far Eastern Republic Revolutionary People's Army A red diamond and a smaller inserted blue diamond on wings and rudder.

The plain red star used from the 1920s to the 2010s, seen on a Polikarpov Po-2.

The Czech Legion Broad red and white stripes across the wings or as a diagonal, bordered with a thin blue stripe.

Sultan of Bokhara's own air force Anti-communist stronghold that acquired three or four ex-RAF Sopwith Strutters. Poor photos show what could be a roundel of red and green based on the sultan's flag, but confirmation is needed.

During the Second World War anti-communist Russians formed a small element of the German forces. There is a possibility that their aircraft wore a white shield, with a blue St Andrews cross on fuselage and wings, and a smaller version on the fin, though this

A Brandenburg of Denikin's Army of the Don Hansa, 1921.
Yefim Gordon

The current Russian red, white and blue star.
via Greg Kozak

marking has to be confirmed. It is far more likely that this emblem was used as a small badge in addition to normal German insignia and/or as a sewn-on badge. Other possibilities were captive aircraft, which had a shield over the swastika or the black German crosses repainted red and the legend 'Frei Deutschland'.

Since 1992 a number of separatist conflicts have occurred within the old Soviet Union. Generally the aircraft involved used the standard red star, but the state of Chechnya used its own insignia, a green and red star with a black and yellow wolf design. There has also been occasional use of a greyed-out version.

2008

RWANDA

This small Central African state, independent since 1962, has used the national flag as a fin flash and a wing roundel of the pan-African colours of red, yellow and green. A new flag of blue, yellow and green was introduced in 2001, and is carried by aircraft, although there are no reports of any roundel version.

SAO TOME E PRINCIPE

These two small islands are a former Portuguese colony in the Atlantic Ocean, independence having been achieved in 1975. Recently a small number of aircraft have been reported with a roundel version of the national flag; this was red, yellow and green, with two black stars, and was mounted on the nose of the aircraft, although this may be a civil identification.

SAUDI ARABIA

A Saudi McDonnell-Douglas F-15.
Anthony Osborne

Hejaz and Nejd were two independent sultanates formed following the defeat of the Ottoman Empire after the First World War. The two states were at war until the formation of the Kingdom of Hejaz-Nejd in 1926, which became Saudi Arabia in 1932. Nejd was the victor in the 1920s war, but seems to have had few, if any, aircraft.

Its national flag was very similar to the present flag of Saudi Arabia. Hejaz obtained a number of aircraft from Italy and the British in Egypt and formed the Royal Hejaz Air Force in 1925. Markings, if any, are unknown, but many aircraft carried Islamic phrases. Hejaz was the originator of the pan-Arab colours of red, black and green.

Originally called the Al Saud Air Arm, the Royal Saudi Air Force was formed in 1933. Aircraft were marked with a roundel version of the flag of Nejd, plain green with a white sword and the legend 'There is one God and Mohamed is his prophet'. The navy carries normal insignia with a superimposed black anchor.

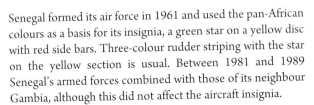

SENEGAL

A Senegal Douglas C-47.
via Kevin Curtis

Senegal formed its air force in 1961 and used the pan-African colours as a basis for its insignia, a green star on a yellow disc with red side bars. Three-colour rudder striping with the star on the yellow section is usual. Between 1981 and 1989 Senegal's armed forces combined with those of its neighbour Gambia, although this did not affect the aircraft insignia.

1913

SERBIA

Serbian military aviation history is divided into two distinct parts: 1912 to 1918, and following the break-up of Yugoslavia in 1992.

Serbian aircraft played a part in both Balkan Wars of 1912 and 1913. At this time the Serbian flag was a horizontal tricolour of red over blue over white, and rudders and wings were marked in this way. During the early part of the First World War a red and blue roundel

A MiG 21 of Serbia-
Montenegro, 2002.
Jozef Grego

A current Soko Super Galeb. Zoltan Rados

1915-17

and rudder marking were used. After the complete occupation by Austria, the French Air Force operated on the Macedonia front, and included several Serbian units whose aircraft carried normal French markings with the Serbian flag on the fuselage.

In 1918 Serbia became part of the Kingdom of Serbs, Croats and Slovenes, and in 1929 part of the federation of Yugoslavia. The federation began to break up in the early 1990s with the independence of Slovenia, Croatia, Bosnia and Macedonia (these countries are dealt with separately). Serbian forces were active in the civil war in Bosnia between the Serbs, Croats and Islamic Bosnians. What was left of Yugoslavia consisted only of Serbia and Montenegro. The roundel was split horizontally as the Yugoslav national flag of red over white over blue. Serbian forces operating in Bosnia used a normal roundel of the same colours. The markings were unchanged in 2003 when what was Yugoslavia became Serbia and Montenegro. By 2006 the colours were changed to the order of the original Serbian flag. Serbia separated from Montenegro in 2007 and adopted a marking similar to that of Yugoslavia from the 1920s to 1941, a white cross on a red and blue roundel with the Serbian order of colours on the fin.

KOSOVO

This part of Serbia, with a large Albanian population, proclaimed de facto independence in 2008. Purchase of aircraft is intended, but no insignia has been decided.

Low-visibility markings on a MiG-29. Zoltan Rados

2013

Coastguard

SEYCHELLES

These islands in the Indian Ocean became independent in 1967. The red, white and green flag dates from 1977, and the air arm from 1990. Aircraft used a roundel formed of the national colours. A new multi-coloured flag was adopted in 1993, and a roundel based on this is now in use.

A Seychelles Aerospatiale Alouette III with a roundel based on the pre-1996 flag.

The current roundel on a Dornier Do 228.
via Greg Kozak

SIERRA LEONE

This West African country gained its independence from Britain in 1961. The military forces have occasionally obtained aircraft. During the civil war of 1991 to 2002 the Sierra Leone Air Wing saw mercenary-led action. Aircraft, usually helicopters, have carried the national colours of green, white and blue either in flag or roundel form.

1968-73

SINGAPORE

Singapore split from the Malaysian Federation in 1968 and soon formed its own air force, the Singapore Air Defence Command, which used a plain red, white and red roundel and fin flash. In October 1973 the markings were changed to a 'yin and yang' symbol, which incorporated a letter 'S'. 1986 saw low-visibility markings for some aircraft, which changed the insignia to a black outline. Officially from 7 November 1990, but normally since

The Singapore plain roundel on a Gloster Meteor F-8. This was a gift from the RAF to Singapore Air Force apprentices. via Greg Kozak

The current low-visibility 'lion's head' marking. via Kevin Curtis

1973-90

The 'S' roundel on a Hawker Hunter T55. via Kevin Curtis

1986-90

January 1991, a new insignia has been used, comprising a red lion's head in a red circle, or, in low-visibility, all in black.

SLOVAKIA

After the occupation of Czechoslovakia by German forces in 1938, the country was split into the Protectorate of Bohemia-Moravia and the Republic of Slovakia. Slovak aircraft were marked with standard German crosses on the fuselage and wings; in addition, the Slovak double cross of blue on a red disc was marked on the wings, adjacent to the German markings, and on the fin. Some Slovak aircraft took part in the invasion of Poland in 1939, during which campaign the Slovak insignia was surrounded by a white ring. In 1941 the Slovak Air Force participated in the German invasion of Russia, and aircraft followed Axis practice by using a cross-type insignia. In this case a German cross type was used, but in blue with a red centre spot, which was marked on wings, fuselage and fin.

With the approach of the Russian armies in 1944, Slovak insurgents set up an air arm with captured German aircraft. These carried the pre-1938 insignia, with a Slovak double cross on the blue portion. The white sector of the roundel was always to the left.

Body, 1938-41

1941-45

Tail, 1938-41

Insurgents, 1944-45

A Slovakian MiG 21MF. Jozef Grego

Czechoslovakia was reunited in 1945, but in 1993 a split was agreed between the Czech and Slovak republics. The Slovak Air Force insignia is now a shield of red and blue and a white double cross.

Low-visibility markings on a Slovakian MiG 29. Jozef Grego

1991-95

SLOVENIA

Although Slovene aircraft took part in the 1919 Carinthia campaign, these are dealt with under Yugoslavia. Slovenia gained the status of an independent republic in 1991, and aircraft of the Slovenian Territorial Defence Force were marked with the white, blue and red of the Slovene arms and national flag. Since 1996 a roundel of white, blue and red has been in use.

A Slovene Pilatus PC9M.

SOMALIA

The Somali Air Force was formed in 1961 and has always used a variation of the national flag, namely a white star on a blue disc. Some aircraft used a light and dark blue fin flash with the white star. For many years there has been no organised military establishment in the country.

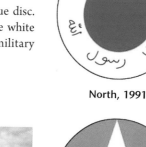

North, 1991

A Somalian MiG 17 providing an unsafe-looking gate guard!

This Hawker Hunter – ex-Abu Dhabi Air Force – looks safer!
via Kevin Curtis

SOMALILAND

On 17 May 1991 the area known as North Somaliland, approximately the previous British Somaliland, broke away from the rest of Somalia. Although unrecognised by the rest of the world, Somaliland set up its own armed forces. The aircraft insignia was originally reported as a white and green roundel, but there is no photographic evidence. A new flag was introduced in 1996, comprising green over white over red with a black star on the white. The Somaliland government lists a roundel of these colours, but that has yet to be confirmed.

Another area that has broken away from Somalia is Puntland, which has two helicopters bearing the flag of Puntland.

Somaliland

SOUTH AFRICA

The South African Aviation Corps was founded in 1915 to assist the army in its conquest of German South West Africa; aircraft wore standard RAF markings. With the fall of the German colony in 1917 the Corps was disbanded.

A South African Spitfire F-XIV, c1948, with normal RAF markings but with orange instead of red.
Stefaan Bouwer

1921-22

1922-24

1924-47

1947-57

1957-94

The South African Air Force was formed on 1 February 1920 and initially used RAF insignia. In 1921 this was changed to a roundel and rudder striping of blue, red, green and orange, which was soon changed to blue, yellow, red and green. From about 1924 the insignia was changed again to the standard RAF pattern, but with the red changed to orange. This continued throughout the Second World War and beyond.

By 1950 a new insignia was applied, first on the wings and eventually on the fuselage. This replaced the central spot with an orange springbok, but the rudder striping remained the same. In 1959 a completely new insignia was devised, which was a representation of Cape Town's fort in blue with the orange springbok. Low-visibility requirements in the 1980s saw a black outlined version in use.

With the change in the political situation in 1994 the rudder striping was dropped and the springbok was replaced by a flying eagle. From about 1995 the new national flag was used as a fin flash. In 2003 the fort was changed to a nine-pointed star, for the nine main regions, in blue with the orange eagle; this is often seen as a black or grey outline.

HOMELANDS

In the 1970s some areas of the country were designated African 'Homelands', with responsibility for their own defence.

Bophuthatswana Formed in 1981, its aircraft carry the coat of arms on the fuselage and occasionally on the wings.

Ciskei Formed in 1982, using the blue and white national flag.

Venda Formed in 1983, using the national emblem on the fuselage and fin.

Transkei Formed in 1986, the few aircraft carried South African civil registrations. It is possible that some carried red, white and green wing stripes.

These were all amalgamated into South Africa on 27 April 1994.

Bophuthatswana, 1981-94

Ciskei, 1982-94

Venda, 1983-94

Transkei, 1986-94

1994

SOUTH SUDAN

This new nation was founded as a breakaway from Sudan in 2011. The few aircraft, usually helicopters, carry the national flag and the arms of the government on a white disc.

A Beech KingAir of South Sudan. The national flag is the only marking so far. Helicopters are also marked with the complex government coat of arms.

SPAIN

Spain's first air force was established in 1911 and saw action in Morocco before the First World War. Aircraft were marked with wings and horizontal rudder stripes in the national colours of red, yellow and red, and by 1918 roundels in these colours were in use.

Spain became a republic in 1931 and the colours, but not the format, were changed to red, yellow and purple. The air arm of the Spanish Navy, which had been established in 1917, used a black anchor on the yellow from about 1922.

The Spanish Civil War broke out in 1936 and aircraft initially carried the same markings. This war, especially on the aviation side, was notable for the large amount of foreign involvement. It therefore became urgent for the opposing aircraft to carry different insignia.

1914-18

REPUBLICAN FORCES

As this was the recognised government of Spain in 1936, its aircraft continued to carry the colours of the republic. The small air arm was

1918-31

1931-36

1936-39

Republic, 1939-42

Republic, 1942

A Spanish Eurofighter Typhoon. Chris Lofting

much enlarged with help from the Soviet Union and France, among others, and these aircraft wore wide red stripes around the wings and fuselage, in addition to, and later in place of, the roundels. They normally retained the three-colour rudder markings.

NATIONALIST FORCES

These aircraft were part of the force that invaded Spain from Morocco; they were largely supplied by Italy and Germany and often flown by pilots of those countries. Aircraft had a white painted rudder with a black saltire cross. Wing and fuselage markings consisted of a black disc often adjacent to three black or white stripes. Wings often featured a white saltire cross, or a smaller cross on the black disc. The fuselage disc often carried a white unit marking.

POST-WAR SPAIN

The Nationalist forces won the war in 1939 and kept to their markings. The stripes and wing crosses were eventually replaced by red, yellow and red roundels, and the black fuselage disc received a 'Falange' clutch of arrows in white or red. From 1942 the roundels placed those on the fuselage and the black cross on a white rudder was retained, and is still in use today.

SRI LANKA

A MiG 15 carrying the original markings when Sri Lanka was called Ceylon.

1950

2009

This island was originally known as Ceylon, and changed its name to Sri Lanka in 1972. Although volunteer units had been formed during the Second World War, the Royal Ceylon Air Force was not established until 1950, and the adopted marking was a red and yellow roundel with side bars of orange and green, and a fin flash in red and orange.

In 2008 a rebel group called the 'Tamil Tigers' obtained some aircraft, but their markings, if any, are unknown.

From 2009 a roundel without bars has been introduced, in red, green and orange with a yellow border. Many aircraft have side bars of varying colours, which are unit emblems similar to earlier RAF squadron markings.

SUDAN

This large African country became independent of joint British and Egyptian rule in 1956 and immediately formed a small air arm. The original national flag of blue over yellow over green was used as a fin marking and, in roundel form, on wings and fuselage. In 1969 the colours were changed to the pan-Arab colours of black, red, white and green, and the national flag of red, white and black with a green segment was adopted; it was used as a fin flash with a roundel form on wings and fuselage.

After a protracted civil war the south of Sudan eventually broke away, and its aircraft are dealt with under South Sudan.

1956-69

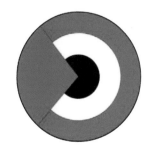

*Sudanese Mil-8.
via Greg Kozak*

SURINAME

The small air force of this former Dutch colony was established in 1983. Initially the national flag was used, but has now been replaced by a roundel form with the flag as a fin marking. Colours are red, white and green with a yellow star.

This CASA 212 Aviocar carries the current roundel. via Greg Kozak

SWAZILAND

1979

Aircraft of the air wing of the Swaziland Defence Force, founded in 1979, carry the national flag, which is blue, yellow and red and bears the shield and spears of royalty.

SWEDEN

Both the Swedish army and navy obtained aircraft in 1911, but the Swedish Air Force was not established until 1926. After the

A Swedish DH 71 Tiger Moth of 1937 with black crown markings. Jan Jorgenson

1914-15

1915-23

1923-37

A Saab J29 carrying the current roundels.
Jan Jorgenson

outbreak of war in August 1914, Sweden began to mark its military aircraft with a blue and yellow wing roundel, and a swallowtail version of the national flag on the fin. A large black letter 'S' was also marked outboard of the roundels. In September 1915 three black crowns replaced the roundels, and by 1917 these were also painted on the fuselage. Initially dark-painted aircraft had the crowns on a white area, but by 1923 white crowns were marked on dark surfaces.

During 1925-26 army aircraft had white crowns with black borders, whereas navy aircraft had black crowns.

With the formation of the air force in July 1926 three black crowns on a white disc was standardised. The flag marking was replaced by blue and yellow rudder striping. In May 1937 the roundel colours were changed to yellow crowns on a blue disc, and the fin markings eliminated. A yellow ring was added to the insignia in 1940. As a neutrality marking, the insignia was considerably enlarged until 1945.

A Saab 340 with low-visibility markings.
Jan Jorgenson

The current marking has reverted to the 1937 version, and greyed-out low-visibility roundels are now frequently used.

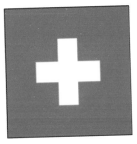

1915-45

SWITZERLAND

A Swiss Dewoitine D.26 with wrap-around markings.

The Swiss Air Force was formally established in July 1914, and by late 1915 its aircraft were marked with the Swiss flag. An area of the wings as large as possible was painted red and marked with a white cross; usually the whole rudder was red with a centrally placed white cross.

During the Second World War neutrality markings of red and white stripes were used in addition to the normal insignia. Since 1945 a roundel form of the flag on the wings and fin has been the usual arrangement.

A McDonnell-Douglas FA 18 Hornet. Jan Jorgenson

1940-45

1945

1949-58

SYRIA

Syria started to obtain aircraft immediately after its independence in 1946, but did not officially establish an air force until 1948. The first markings were a green, white and black roundel and horizontal fin flash with three red stars on the white.

1958-61

1963

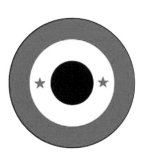

A Syrian Gloster Meteor F8. via Greg Kozak

In 1958 Syria joined with Egypt and Yemen to form the United Arab Republic, and roundels and flag were changed to red, white and black with two green stars. In 1961 Syria left the union and reverted to the 1948 markings.

1963 saw the revival of the 1958 version, but with three stars. In 1971 the stars were dropped altogether, and in 1980 the present flag and roundels brought back two stars.

Rebel forces do not have any aircraft as yet; they have used the flag of 1948.

TAIWAN

After the Chinese Civil War of 1949 the remnants of Chiang Kai-shek's nationalist air force escaped to the island of Formosa – Taiwan in Chinese. Aircraft have continued to use the national markings for all of China prior to 1948, which comprises a white twelve-pointed star on a blue disc, and a rudder marking of blue and white stripes.

As from 1991 Taiwan is no longer a one-party state. Since then roundels have been modified to differentiate between the state, which included the armed forces, and the National Party; this has been done by using a smaller white star. Greyed-out low-visibility markings are also in use.

A Taiwanese Lockheed F-104G Starfighter.
Andreas Zeitler

1949-91

TAJIKISTAN

Independent since 1991, this former Soviet Central Asian republic has obtained a number of military aircraft. Those seen have carried the national flag of red, green and white with the addition of, or instead of, fuselage stripes in these colours.

TANZANIA

The government of Tanganyika possessed a number of aircraft that carried the national flag of green, black and green, separated with a thin yellow line.

Tanganyika and Zanzibar united to form Tanzania in 1964, and an air wing of the People's Defence Force was soon established. Its aircraft have always carried the same roundel, a yellow torch and wreath on a green disc with a blue border. It appears that combat aircraft carried no markings and only recently bore the emblem of the air force on the fin.

A Tanzanian Piaggio FWP149. George Trussell

THAILAND

1917-41

Officers of the Siamese Army went to France for flying training in 1911, and in March 1914 the Royal Siamese Flying Corps was formed. Siamese officers took the Allied side in the First World War, which resulted in the adoption of the current red, white and blue national flag in 1917. Aircraft based in Siam were marked with a roundel version of the flag, and rudders were striped according to its proportions.

The corps became the Royal Siamese Air Force in 1937. In 1939 the country changed its name to Thailand and the force became the Royal Thai Air Force. In January 1941 war broke out between Thailand and the French in Indo-China. Because of a similarity of markings, the wings of Thai aircraft carried the actual

1941-42

Left: *A Thai Navy Cessna U-17.* via Kevin Curtis

Right: *A Thai Fairchild Provider C-123.* via Kevin Curtis

Tail, 1942-45   **Wing, 1942-45**

national flag. Japan invaded in December 1941, and after a two-day war the country became, in effect, an unwilling satellite of Japan. Aircraft of this period were marked with a combination of Japanese roundels and the pre-1917 flag of a running white elephant on a red field; the latter was carried on the fin and under the wings.

After the Japanese defeat a return was made to the pre-1941 roundel, and rudder striping and later a fin flash were also used. During the 1950s some Thai Navy aircraft had a black anchor on the fuselage side. Modern naval aircraft have a red disc bearing a white elephant superimposed on the national flag fin marking.

TOGO

Togo's air force was founded in 1960 and has always used the pan-African colours of red, yellow and green. The national flag has two yellow and three green stripes with a corner in red bearing a white star; the aircraft insignia uses a roundel version.

A Togo Aerospatiale Super Puma.

TONGA

Tonga is a group of islands in the South Pacific Ocean. The Defence Services Air Wing obtained some aircraft in 1997, and they were marked with a red Maltese cross with a yellow border, surmounted by a royal crown.

TRANSNISTRIA

Although not recognised by any other country, this part of Moldova declared its independence in 1992. A short war with Moldova, assisted by Russia, produced a stalemate. In 2007 Transnistria obtained a number of helicopters, which were marked with a roundel of the national flag, basically red and green.

TRINIDAD AND TOBAGO

2012

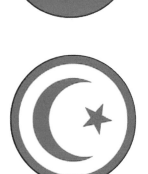

A Trinidad and Tobago Defence Force Piper A31, showing the original markings as national colours in stripes.

A Trinidad Agusta-Westland AW-139 with the latest roundel.

The defence force of these islands in the West Indies was founded in 1966. The national colours are red, white and black, and stripes in these colours have been used, since 2007 in roundel form. Since 2011 the Air Guard has added a yellow bird to the centre spot.

TUNISIA

The Tunisian Air Force was established in 1960, since when aircraft have always carried a red-bordered white disc bearing a red star and crescent.

A Tunisian Aermacchi MB-326.

TURKEY

During the 1912 and 1913 Balkan Wars, Turkey possessed a number of aircraft, mostly flown by mercenaries. They had rudders and a large section of the wings painted red, bearing a white star and crescent.

Germany supplied most of Turkey's aircraft and pilots during the First World War, and by 1916 these carried large black squares on the wings and fins, outlined in white; this was either to bring Turkish markings into line with other aircraft of the Central Powers or because of a shortage of red paint. After the war there was some use of red/white/red roundels.

By the 1922 war with Greece, Turkish aircraft were marked with white-bordered red squares, and red rudders marked with the white star and crescent. Apart from the eventual use of fin flashes, this remained the same until 1972. As the speed of aircraft increased it became difficult to differentiate between the red squares and the Soviet red stars, so a change was made to red and white roundels, which since the 1980s have been reduced in size.

A Turkish Gotha WD13 with 'black square' markings.

1914 insignia on a Gotha WD-2, 1914.

A Turkish Republic F-84F Thunderstreak. via Kevin Curtis

This Northrop T-38 carries the current roundel. Andreas Zeitler

1913-15

1915-18

1918-72

1918

TURKMENISTAN

2007

The national flag of this Central Asian republic was first flown in 1992, and an air arm was formed in 1995 using markings based on it, comprising a white crescent and five stars on a green disc with a red and yellow border. This was modified in 2011 to an eight-pointed border.

2008

2012

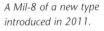

A Mil-8 of a new type introduced in 2011.

A Sukhoi Su-25 with current-style markings.
Jozef Grego

UGANDA

1962-64

Independence for this East African country came in 1962 when a Police Air Wing was formed. Aircraft carried the national flag, a number of red, yellow and black horizontal stripes with, in the centre, a white disc bearing an illustration of the crested crane, Uganda's national emblem. An air force was established in 1964 and aircraft have been marked with a variety of roundels based on the national flag.

A MiG 21 carrying one of the many variations of the Uganda roundel.

UKRAINE

This country seized its independence as the Ukrainian People's Republic in the wake of the Russian Revolution in 1917. German forces invaded Ukraine in April 1918 and set up a puppet state, the West Ukrainian People's Republic. The two Ukraines fought each other and, after the German defeat, were in conflict with the new Soviet Union, the White Russians and Poland. Known markings were all blue and yellow and are taken from the colours of the old Austrian

Republic, 1918

West Ukrainian Republic,
1918-19

West Ukrainian Republic,
1919

Republic, 1919

Republic, 1919-20

province of Galicia, which covered most of what became Ukraine

By early 1919 Ukraine had adopted an ancient and traditional Ataman Cossack trident design, the 'tryzub', which was used in various forms, e.g. plain black, plain yellow, yellow on a blue square. The Ukrainian national flag of blue over yellow was also used, sometimes in chequerboard form. The West Ukrainian Air Force used the same colours as a roundel, and even a yellow skull and crossbones on a blue-bordered white disc. Some, presumably naval, units bore a superimposed black anchor.

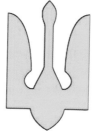

Republic, 1919-20

Republic, 1920

1919-21

Tail

1991-92

1992

Navy

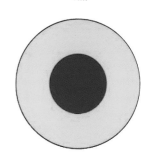

The Red Ukrainian Air Fleet was formed in February 1919, but became part of the Soviet Air Force in May and probably carried a red star. Ukraine was partitioned in 1921 between the Soviet Union and Poland. In 1922 it became one of the constituent republics within the USSR, and received a separate seat at the United Nations in 1946.

With the break-up of the USSR in 1991 Ukraine once more achieved independence. Aircraft now wore blue and yellow roundels, and the trident in yellow on a blue shield became the standard fin marking. There are some reports of a chevron design, but this has never been used. Some aircraft of the Ukrainian Navy have a black anchor on the roundel and helicopters carry the naval flag under the fuselage. Likewise the army helicopters carry the army flag.

A Ukraine Su-27.
Andreas Zeitler

UNITED ARAB EMIRATES

The United Arab Emirates were once known as the Trucial States, and gained their present name in December 1971. Abu Dhabi had formed its air force in 1968, and its aircraft were marked with a red, white and sand-coloured roundel, the red area carrying 'Abu Dhabi' in black Arabic script. The red and white national flag appeared as a fin flash.

Dubai created its Police Air Wing in 1971, and initially the aircraft carried the arms of the sheikdom in red on a white square as a fuselage marking. In 1974 this changed to a red and white roundel with the arms in the centre. The red and white flag of Dubai was used as a fin flash.

1968-76

An Abu Dhabi Hawker Hunter flying for the United Arab Emirates.
via Kevin Curtis

1971-74

From the formation of the joint command of the Emirates in 1976 the fin flash was changed to that of the United Arab Emirates, which dates from 1972. Although Abu Dhabi and Dubai have some autonomy, all aircraft now bear the UAE insignia. Sharjah, the only other emirate with its own forces, formed an air wing in 1984, but its aircraft have always carried the UAE insignia, a roundel of green, white and black with a red segment across the green and white.

1974-76

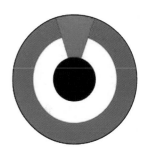

A Lockheed-General Dynamics F-16.
Andreas Zeitler

1914-15

1914

Tail, 1915-37

UNITED KINGDOM

The Royal Engineers formed the first Air Battalion in 1911, then in May 1912 the Royal Flying Corps was established with army and navy sections. By 1913 the navy had formed its own Royal Naval Air Service, and in April 1918 these two forces joined together to form the Royal Air Force.

Early aircraft marked their ownership with a Union flag painted on the rudder, and it was with this marking that the first aircraft landed in France, in August 1914, on the outbreak of war. On 26 October 1914 an instruction was issued to mark wings, fuselage and fin with the Union flag. By late 1914 it became obvious that shape rather than colour was the dominating factor in differentiating friend from foe, so it was decided to adopt the French system of a three-colour roundel. From 11 December 1914 aircraft would carry a roundel with the colours in reverse order from the French, although naval aircraft often used the French order to distinguish them from the Flying Corps; they also continued to use the Union flag on the rudder.

Officially, from 16 May 1915 the naval service adopted a red and white roundel, but this was only temporary. By early 1916 all British aircraft carried roundels with a red centre, and rudder striping with the blue leading. This marking continued with very little change until 1937, a notable exception being the adoption of red and blue roundels for night use.

The increasing tension in Europe by 1937 saw the introduction of camouflage to aircraft. Initially roundels were outlined in yellow, but this soon gave way to red and blue roundels above the wings and yellow outlines on the fuselage only. After the outbreak of war in 1939 markings were standardised to a normal roundel under the wings, a red and blue roundel above, a normal roundel outlined in yellow on the fuselage, and a red, white and blue fin flash, the red leading.

A Bristol Scout 'C' in 1914, with the national flag on the wings and fuselage.

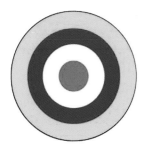

1915-37

Body, 1937-42

Above: *An RAF Hawker Hurricane from the early Second World War.* Author

Right: *A De Havilland Mosquito from later in the war.* Author

A Fairchild Argos, bearing the two-tone blue roundels of South East Asia Command. Author

Tail, 1937-42

Wing, 1937-42

South East Asia, body, 1942-45

South East Asia, tail, 1942-45

Body, 1942-47

South East Asia, 1942-47

A Chance-Vought Corsair, South East Asia Command, 1944. Author

Tail, 1942-47

Wing, 1942-47

1047-72

Tail, 1972-85

A De Havilland Chipmunk with the 1960s-type roundel. Author

A Westland Lynx of the Royal Navy with a red/blue roundel. Author

From July 1942 the yellow surrounds and the white portion of the markings were considerably reduced in size, and except for aircraft operating in the Far East this system was maintained until 1947. During the Japanese invasion of 1941-42 aircraft wore standard insignia. South East Asia Command was formed in 1942 to integrate all Allied units, and its markings were a roundel and fin flash of dark blue and light blue in a smaller size. Some Royal Navy squadrons in this theatre used roundels of various designs, but always with the red eliminated. Some US-supplied aircraft retained the American blue and white side bars.

Current markings in red, white and blue date from 1947. The roundel differed from earlier designs by having a larger central red spot. In the early 1970s experiments resulted in smaller red and blue markings, and since the 1980s these have often been painted in toned-down paler shades.

A British Aerospace Nimrod with pale low-visibility roundels. Author

1972-85

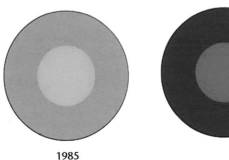

1985

UNITED NATIONS

The United Nations charter was signed in 1945, and the flag of a white laurel wreath and globe on a blue background was adopted on 20 October 1947. It has always been one of the main points of the charter that member countries would supply military equipment, as necessary, to restore order or bring help to troubled areas around the world. The first instance was in the aftermath of the Israeli War of Independence in 1948.

UN aircraft are normally painted white overall and carry, in large black lettering, 'UNITED NATIONS' or 'UN', and the flag as a fin flash. Camouflaged aircraft carry the lettering on a white or pale blue panel. There has been some use of the United Nations emblem as a wing roundel, and recently an increasing use of the national flag of the donor country, usually marked on the fin and smaller than the UN flag.

UNITED STATES

The Aeronautical Division of the US Army Signal Corps was established in August 1907, but progress was slow, and the army and navy owned very few aircraft by 1914. Until 1917 ownership of aircraft was established by a rudder and wings insignia comprising a red five-pointed star for the army and a blue anchor for the navy. From May 1917 regulations on insignia for both services were standardised on a wing marking of a white star on a blue disc with a red centre spot, while rudders were striped in blue, white and red, the blue leading. This marking continued in use during and after the First World War.

1914-17

An American De Havilland DH 9 with the red star only on fin 1914.
via Greg Kozak

A Nieuport 28 of the American Expeditionary Force, 1918.
via Greg Kozak

1917-27

A Boeing-Stearman Kaydet, 1936. Jan Jorgenson

AEF, 1918-19

1942-43

1943-47

1943

1947

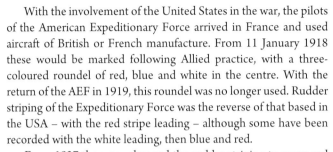

With the involvement of the United States in the war, the pilots of the American Expeditionary Force arrived in France and used aircraft of British or French manufacture. From 11 January 1918 these would be marked following Allied practice, with a three-coloured roundel of red, blue and white in the centre. With the return of the AEF in 1919, this roundel was no longer used. Rudder striping of the Expeditionary Force was the reverse of that based in the USA – with the red stripe leading – although some have been recorded with the white leading, then blue and red.

From 1927 the army changed the rudder striping to seven red and six white horizontal stripes with a vertical blue stripe next to the hinge. Naval aircraft were devoid of rudder markings. In 1941 the navy followed the army with the rudder markings until late 1942 and a brief period from 1946 to 1947. All rudder markings were discontinued between 1942 and 1946. Following Allied practice, American aircraft eliminated the red part of the insignia in all theatres of war. A plain white star on a blue disc above the port and below the starboard wing was in use from May 1942. Aircraft participating in Operation 'Torch', the invasion of North Africa, outlined their markings in yellow. By 1943 it was apparent that shape rather than colour was a deciding factor in aircraft recognition, especially true in the Pacific area. Bars were therefore added to the roundel and for a brief period the whole insignia was outlined in red; by the end of 1943 a blue line was standardised. The basic shape and positioning of insignia was normal for other Allied aircraft operating in the region. A red stripe to the bar began to appear from 1945, but became official on 16 June 1947 and was universal for all military aircraft on the founding of the United States Air Force on 26 September 1947.

A North American P-51 Mustang, 1945. Author

During the war in Vietnam there was a requirement for a low-visibility insignia, which was achieved by a considerable reduction in size. The current trend has meant that the insignia has changed to a pale grey colour or even a simple black outline.

Several other American organisations have used special aircraft markings, two of which are more military than civilian. The US Marines had their own insignia during the First World War, comprising an anchor over the three-colour roundel. The Coast Guard from the late 1930s to the early 1940s marked the rudders of its aircraft with the top third in red and the rest in three vertical blue stripes, while across the wings were the letters 'USCG'.

A Douglas TA-4J Skyhawk, 1980. via Kevin Curtis

A Bell Iroquois TH-1H with low-visibility markings.

URUGUAY

The original flag of Uruguay, until 1828, was a horizontal tricolour of blue, white and blue with a broad red diagonal band, and rudder striping and a roundel version of this flag were used on Uruguayan military aircraft following the formation of the air force in 1916.

In 1924 a naval arm was formed, which used the modern Uruguayan flag as a rudder marking; this comprised nine blue and white horizontal lines, representing the original nine provinces, with a white quarter containing the independence emblem, the yellow 'Sun of May'. These aircraft also carried black or white anchors inboard of the air force roundel.

1916

*A Uruguayan Navy
Grumman Hellcat.*
Santiago Rivas

Navy

Since 1954, in common with many Latin American countries, Uruguay has applied its wing insignia above the port and below the starboard wing. From the 1990s a fin flash rather than a rudder marking has been in use.

UZBEKISTAN

The national colours of blue, white and green in roundel form, with a thin red line between the colours, have formed the air force insignia of this Central Asian republic since independence in 1991. This colour scheme is often striped across the fin.

*An Uzbekistan Sukhoi
Su-27.* via Greg Kozak

VENEZUELA

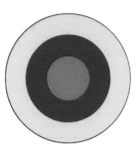

1920-56

Venezuela's air force was established in 1920 and used the national flag, a horizontal tricolour of yellow, blue and red, as rudder striping and roundels. Since about 1956 bars in the three colours have been added to the roundels, and it has been normal practice to display them above the port and below the starboard wings. The blue central area of the rudder marking bears a semi-circle of seven white stars for the original seven provinces. Low-visibility markings have been reduced in size and rudder markings are rarely seen. Some aircraft now wear just the national flag on the fuselage. Naval aircraft have reverted to the pre-1956 roundel and a black anchor on a white square.

A Venezuelan IAI Arava.

A Grumman Tracker. via Kevin Curtis

A North American T-6. George Trussell

VIETNAM

Tail, 1950-54

Once part of French Indo-China, Vietnam formed its first air arm, under French supervision, in 1950. Aircraft were marked with an orange disc bearing three concentric red circles, and the fin marking was in rectangular form, comprising the national flag.

After the French withdrawal in 1954, the country was divided into North and South. The Southern Zone began to receive aid from the United States and, by 1962, the wing and fuselage

A pre-1954 Vietnamese MS500 (French-built Fiesler Storch). via Greg Kozak

1950-54

markings were changed to a US type, although the fin marking remained unchanged. The new roundel was a white star on a blue disc with side bars of orange with a red stripe. The whole insignia was surrounded by a red border.

The Northern Zone adopted a national flag of plain red with a yellow five-pointed star. The initial aircraft marking was a yellow star with a red border, but some have been reported carrying a plain yellow star, often as a fin marking. Some captured American aircraft carried the Viet Cong flag, red over blue with a yellow star. By about 1970 North Vietnam placed the yellow star on a red disc with red side bars, all surrounded with yellow. Unified Vietnam, from 1975, used this North Vietnamese marking or the national flag.

A South Vietnamese DHC-2. via Kevin Curtis

North Vietnam, 1951-65

A North Vietnamese MiG 21 with markings now current for all Vietnam.

South Vietnam, 1962-75

YEMEN

A tentative attempt was made to form an air arm in 1926, the aircraft bearing Arabic inscriptions on red or possibly green panels on the fuselage and fin. It was not until 1955 that any viable military aviation unit was established. This was, in effect, the Imam of Yemen's private air fleet and carried his personal emblem, a red flag bearing a white sword and five white stars. A photograph is available of an aircraft that seems to have a fuselage marking of a roundel version of this emblem.

1956-62

1962-75

Between 1958 and 1961 Yemen had a loose association with the United Arab Republic. With the establishment of the Yemen Arab Republic in 1962 a new flag, red over white over black with a single green star, was adopted, and this formed the basis for the insignia of the Yemeni Air Force. The green star was dropped by 1979 and, for a brief period, the national flag, and therefore the fin flash, featured white crossed swords on the red area.

A Lockheed C-130 with the pre-1975 green star. via Kevin Curtis

SOUTH YEMEN

In 1962 the British-controlled Aden Protectorate became the Federation of South Arabia, and its flag was dark blue over green over blue, separated by thin yellow lines, with a white star and crescent superimposed. Aircraft of the South Arabian Air Force carried the national flag, without the star and crescent, as a fin flash. A roundel in the national colours, split horizontally and including the star and crescent, was used as a wing and fuselage marking.

The British left Aden in 1967 and the country became known as the Yemen People's Democratic Republic. The flag was the Yemen Republic's tricolour but with a blue triangle with a red star at the hoist. This was used as a fin flash, while the wing and fuselage insignia was a blue triangle with a black border and a centrally placed red star. By 1980 this had been changed to a three-colour roundel with a blue segment and a red star at the top.

Although not fully established until 1994, the unification of North and South Yemen was agreed in 1990. The new unified Yemen had a simple red, white and black roundel and fin flash.

Tail, South Arabia, 1963-67

South Arabia, 1963-67

South Yemen, 1967-80

South Yemen, 1980-90

A South Arabian Federation DHC-2, 1963.

A Yemen People's Democratic Republic (South Yemen) British Aerospace Strikemaster, 1968. via Kevin Curtis

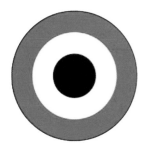

A Yemeni MiG 21.

YUGOSLAVIA

In 1918 the Kingdom of Serbs, Croats and Slovenes was formed from the kingdoms of Serbia and Montenegro and the Austrian provinces of Croatia, Slovenia and Bosnia-Herzegovina. Its first flag was blue over white over red. Former Austrian aircraft, usually flown by Slovenes, saw action over Carinthia during 1918-19, and carried the new kingdom's flag as a rudder marking and the national colours in varying forms as wing and fuselage insignia. One type was roundels with the colours in various orders, but most settled on blue/white/red. Chevron designs were also used.

By the later 1920s a new roundel in white with a blue border on a blue and white roundel was in use. The country's name was changed to Yugoslavia in 1929 and the roundel changed to the cross over a red, white and blue roundel.

Germany occupied the country in 1941 but strong partisan forces continued to fight on, and by 1943 they were being supplied with aircraft from the Allies. Markings for these tended to be RAF with a red star painted over the roundel and added to the area of the fin. The partisans captured many German aircraft and these were flown with the markings painted out and a red star substituted. After the war a red star, sometimes bordered in yellow, on a blue and white disc was applied, the horizontal fin flash having the red star.

After the break-up of Yugoslavia in the early 1990s the red star was dropped and a new roundel of the national flag split horizontally was adopted as a roundel. The markings remained unchanged when Yugoslavia became Serbia and Montenegro in 2003.

For further information see Bosnia, Croatia, Macedonia, Montenegro, Serbia and Slovenia.

1918-22

1918-29

1922-29

1929-41

Insurgents, 1941-45

1945-92

1992

A Balkan Air Force North American T-6 of 1945.
Vladimir Ristic

A Soko Jastreb in pre-1992 markings (for post-1992 see Serbia). Zoltan Rados

Tail

ZAMBIA

Formerly Northern Rhodesia, Zambia became independent in 1964 and adopted a national flag of green, red, black and orange, the orange symbolising the importance of copper in the nation's economy. The fin flash and roundels are based on the national flag, the roundels also featuring a white eagle.

A Zambian Aermacchi MB-326.
Winston Brent

ZIMBABWE

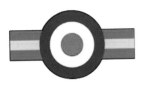

Southern Rhodesia, 1947-53

1953-63

1963-75

Before 1980 Zimbabwe was known as Rhodesia, and earlier as Southern Rhodesia. The Southern Rhodesia Air Unit was formed in 1937 and became the Air Force in 1939. Aircraft carried standard RAF markings but with serial numbers beginning SR. From 1947 they were distinguished by the addition of green/yellow/green side bars to the fuselage roundels.

The Central African Federation of Rhodesia and Nyasaland was formed in 1953, and aircraft of what was now the Royal Rhodesian Air Force carried standard RAF roundels with three small assegai spears on the central spot. With the break-up of the Federation in 1963 the markings bore a single large assegai across the roundels.

Rhodesia declared independence in 1965 but continued to use the same markings until the late 1970s. The new insignia adopted was a green and white fuselage marking, the white area bearing a golden lion, and a green and white fin flash. There were no wing markings. By 1979 all national markings were deleted from aircraft on active service.

Above: *A Rhodesian Percival Pembroke in 1950.* via Kevin Curtis

A Rhodesian Percival Provost bearing the three assegai markings, 1960. via Greg Kozak

*A Rhodesian DH Vampire
T.11 with the single
assegai insignia, 1965.*
via Greg Kozak

1975-79

Rhodesia became Zimbabwe in 1980, and initially no markings were carried. In 1982, however, aircraft carried a yellow representation of the soapstone bird, the national emblem, on the fin. In 1994 a roundel of green, yellow, black and white was marked on the fuselage and wings. The national flag was carried on the port side of the fin, the soapstone bird on the starboard.

1982-94

*The single assegai marking on a
Rhodesian Canberra.* Winston Brent

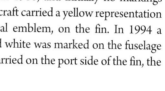

*A Rhodesian DH
Vampire, 1976.*

*A Zimbabwean
CNAMC-K8.*

BIBLIOGRAPHY

Andersson, L. *A History of Chinese Aviation until 1949* (2008)

Apostolo, G. and Bignozzi, G. *Warbirds: Military aircraft of WW1 in colour* (1973)

Brent, W. *African Air Forces* (1999)

Cooper, T. et al *African MiGs* (Vol 1 2010, Vol 2 2011)

Cornejo, H. *History of Mexican Air Force*

Dienst, J. and Hagedorn, D. *Latin American F-51 Mustangs*

Donald, David and Lake, Jon *Encyclopaedia of World Military Aircraft* (1996)

Flintham, V. *Air Wars and Aircraft: A detailed record of air combat 1945-* (1989)

Gordon, Y. *German Aircraft in Soviet Union and Russia* (2008)
    MiG 17 (2002)

Gordon, Y. and Komissarov, D. *Soviet and Russian Military Aircraft in Africa* (2013)
    Soviet and Russian Military Aircraft in the Middle East (2013)

Green, W. *Flying Colours* (1981)
    Air Forces of the World (1958)
    World Air Power Guide (1963)

Hagedorn, D. *North American T-6* (2009)
    Central American and Caribbean Air Forces (1993)
    Latin American Air Wars (2006)

Hagedorn, D. and Hellstrom, L. *Foreign Invaders* (1994)

Krivinyi, N. *World Military Aviation* (1977)

Nedliakov, D. *Aircraft of the Kingdom of Bulgaria* (2001)
    Aircraft Manufacture in Bulgaria (2009)

Neulen, H. W. *In the Skies of Europe* (1998)

Peacock, L. *The World's Air Forces* (1991)

Robertson, B. *Aircraft Camouflage and Markings 1912-1954* (1956)
    Aircraft Markings of the World 1912-1967 (1967)

Stafrace, Charles *Arab Air Forces* (1994)

Stapfer, H. *Warsaw Pact Air Forces* (1991)

Steenkamp, W. *Aircraft of the South African Air Force* (1981)

Talbot-Booth, E. *Aircraft of the World* (c1935)

Taylor, M. *Encyclopaedia of World Air Forces* (1988)

Wall charts published by, among others, *Air Forces Monthly, Avions* (French), *Air Progress* (Canada c1938), *National Geographic* (US c1934), Small Air Forces Clearing House, and official RAF and Luftwaffe Second World War publications.

Magazines including those published by IPMS in the UK, France, Belgium, Australia, Greece, Finland, Chile, etc., *Air Britain, Air International, Air Enthusiast, Air Forces Monthly, Air Pictorial, Air Zone* (France), *Avions* (France), *Flieger Revue* (Germany), *Flight, Insignia, Small Air Forces Observer, Scale Aircraft Modelling, Scale Aviation Modeller, Wings of Fame, World Air Power Journal,* and many others throughout the world.

Organisations including Small Air Forces Clearing House, Air Combat Aviation Group, Linden Hill Decals, Blue Rider, and many websites too numerous to mention, together with the many branches of the International Plastic Modelling Society worldwide.

Individuals Far too many to mention individually, but mention must be made of the members of SAFCH, Dan Hagedorn, Lennart Andersson, Greg Kozak, Jozef Grego, and enthusiasts from Russia, Latin America, Japan, China, Australia, South Africa and elsewhere.

INDEX OF COUNTRIES

INDEX OF AIRCRAFT

Luftwaffe
EMBLEMS
1939-1945

Barry Ketley

One of the most striking features of Luftwaffe aircraft during the Second World War was the widespread use of distinctive badges and emblems that identified individual Luftwaffe units. Some of these insignia remained in use throughout the war and even into the final days of the conflict.

The Luftwaffe units using their own insignia ranged in size from full-blown *Luftflotten* (Air Fleets) down to small temporary groupings of a *rotte* (pair) of aircraft. Many such units were famous fighter and bomber units – such as *JG 2, ZG 76, KG 2* and *KG 54*. Others were less well-known including maritime squadrons, humble training units and even communication flights. Whatever their status, they proudly carried their own unique badge on their aircraft, but until now there has been no comprehensive reference source for these emblems.

Now, this fully revised and greatly expanded edition of *Luftwaffe Emblems* provides a unique reference to these markings and features over 1,000 emblems – many of which were previously unknown – along with many previously unpublished photographs. Grouped by squadron type and function, *Luftwaffe Emblems* is an exceptional reference aid for modellers, historians, enthusiasts and anyone with an interest in WWII Luftwaffe aircraft.

144 pages softcover
268mm x 198mm
Over 1000 colour artworks
Over 120 photographs
9780955 426834 £14.95

Order online at www.crecy.co.uk or from any bookshop